Wastewater Reuse 2

Wastewater Reuse 2

Micropollutants, Membranes and Treatment Procedures

Kader Gaid

iSTE

WILEY

First published 2025 in Great Britain and the United States by ISTE Ltd and John Wiley & Sons, Inc.

ISTE Ltd
27-37 St George's Road
London SW19 4EU
UK

www.iste.co.uk

John Wiley & Sons, Inc.
111 River Street
Hoboken, NJ 07030
USA

www.wiley.com

Library of Congress Control Number: 2024950901

British Library Cataloguing-in-Publication Data
A CIP record for this book is available from the British Library
ISBN 978-1-78630-980-8

Contents

1

Case Studies with Tertiary Treatment

The effective integration of wastewater reuse into water resource management is based on the quantity of water required for a specific application, as well as the associated water quality requirements.

The key challenge lies in choosing the most appropriate process to meet local constraints. The technical tools available are those already used for wastewater or drinking water treatment.

The process of selecting a tertiary treatment system involves at least three steps, the first of which is to assess the performance of available processes in relation to the quality of the water to be treated and the end use. The next step is to analyze regulatory constraints. The techno-economic assessment takes into account existing infrastructure and equipment, as well as sources of financing for investment and operation.

The quality requirements for each use are not the same. Generally speaking, non-drinking uses only require minor treatment.

A significant number of tertiary filtration-disinfection plants have been built for urban water reuse applications such as parks, school grounds and golf courses, as these processes involving solid/liquid separation and disinfection are simple and inexpensive.

There are significant regional and seasonal variations in water use models. For example, in urban areas, industrial, commercial and non-drinkable urban water requirements are the main demand for water. For agricultural applications in arid and semi-arid regions, irrigation is the main reason for water demand. Water

---

For a color version of all figures in this chapter, see www.iste.co.uk/gaid/reuse2.zip.

requirements for irrigation applications tend to vary seasonally, while industrial water needs are more constant.

Tertiary treatment can therefore be used for a variety of purposes, as summarized in Table 1.1.

Use categories	Uses
Non-drinking, urban	Water for public parks, sports facilities, private gardens, roadsides, street cleaning, fire hydrants, vehicle washing, air conditioning and dust control
Agricultural uses	Non-commercially processed food crops, commercially processed food crops, dairy pastures, forage, fibers, seed crops, ornamental flowers, orchards, hydroponics, aquaculture, greenhouses and viticulture
Industrial uses	Cooling water, recirculating cooling towers, washing water, washed aggregates, concrete making, soil compaction and dust control
Recreational areas	Golf courses, recreational ponds with/without public access (e.g. fishing, canoeing and swimming) and aesthetic reservoirs without public access
Environmental areas	Wetlands, swamps, marshes, flow augmentation, wildlife habitat and forestry

Table 1.1. *Uses with tertiary treatment*

The choice of tertiary treatment includes the following processes:

– membranes included in secondary treatment (membrane bioreactor (MBR));

– simple disinfection: chlorine, peracetic acid, performic acid;

– combined disinfection: ozone-chlorine, UV-chlorine;

– tertiary filtration-disinfection: chlorine tertiary filtration, ozone-chlorine tertiary filtration, UV-chlorine tertiary filtration;

– quaternary treatment: ozone-granular activated carbon, advanced chemical oxidation;

– microfiltration (MF) membranes, ultrafiltration (UF) membranes, MF or UF followed by nanofiltration (NF) or reverse osmosis (RO) membranes;

– MF and UF can be coupled with advanced oxidation, ozone-GAC, etc.

The choice of process depends on a number of factors, the most important of which are effluent quality, type of use, quality requirements and plant size.

Depending on local conditions and technical and economic criteria, different processes can be considered.

Combined filtration-disinfection technologies or MF membranes guarantee improved water quality for the various uses listed in Table 1.1. Their implementation is helping to ensure better environmental protection and new applications in urban areas.

They reliably ensure good water quality (removal of suspended solids, viruses and pathogens). The most common method of disinfecting wastewater is chlorination. In some countries, the potential toxicity of chlorine by-products makes this technique less appealing. Ultraviolet irradiation appears to be an emerging method. Its efficacy is comparable, if not often superior, to the elimination of viruses and bacteria, but it has no residual. Ozone disinfection is not widely used, but it is gaining ground due to its excellent effectiveness in inactivating pathogens.

Tertiary treatment therefore allows direct reuse, in other words, the immediate use of treated wastewater, without passing it through or diluting it in the natural environment.

Due to consumer quality requirements, this treatment can only be applied for non-drinkable uses in the agricultural (irrigation), industrial and urban sectors.

The microbiological guarantees expected for each use are summarized in Table 1.2.

Uses	Reduction of viruses Log 10	Reduction of protozoa Log 10	Reduction of bacteria Log 10
Irrigation of food crops	7.0	6.5	7.0
Car washing	5.5	4.5	5.5
Laundry	4.5	3.5	4.5
Cooling towers	4.5	4.0	4.5
Irrigation of non-food crops	3.5	2.5	3.5
Recreation and leisure areas	4.0	3.5	4.0
Toilet flushing	5.5	5.0	5.5

Table 1.2. *Reduction across the entire chain: from raw water to water for reuse*

Treatment systems must contain both design and operating requirements, so as to ensure treatment reliability. Reliability features, such as alarm systems, emergency power supplies, duplication of treatment processes, emergency storage or disposal of inadequately treated wastewater, monitoring devices and automatic controllers, are important. From a public health perspective, measures must be taken to ensure reliable disinfection.

1.1. Limited urban use, irrigation of non-food or processed crops[1]

These applications are relatively common examples of water reuse in areas of restricted access or activity. Typical examples of this category are as follows: landscape irrigation (golf courses, cemeteries, road verges and highway medians), restricted recreational uses (recreational fishing, canoeing and other non-contact recreational activities), agricultural irrigation (forages, fibers, grain crops, pastures, nurseries, sod farms and commercial aquaculture). Water quality requirements are the same for all uses in this category.

1.1.1. *Irrigation*

The aim is to reduce the number of pathogens in treated wastewater by installing several barriers designed to reduce the likelihood of microorganisms being present.

The main health risk control measures for irrigation with treated and reused wastewater include the following:

– wastewater treatment and quality control during distribution and in storage facilities;

– the choice of appropriate irrigation methods and cultivation practices, such as irrigation of restricted areas (such as freeway landscapes, other areas where the public has limited access or exposure to reclaimed water) and irrigation in areas with free public access (such as golf courses, parks, playgrounds, schoolyards and residential landscape);

– public access conditions (recreational areas, parks, green spaces, etc.);

– other measures, including control of human exposure (e.g. protective measures for farm workers such as gloves and masks), harvesting measures, education of the public concerned and promotion of good hygiene practices.

1 Some of the technical and scientific descriptions presented in this book have been partly taken from *Drinking Water Treatment 4 and 5* (Gaid 2023), as the processes used in the two applications are very similar.

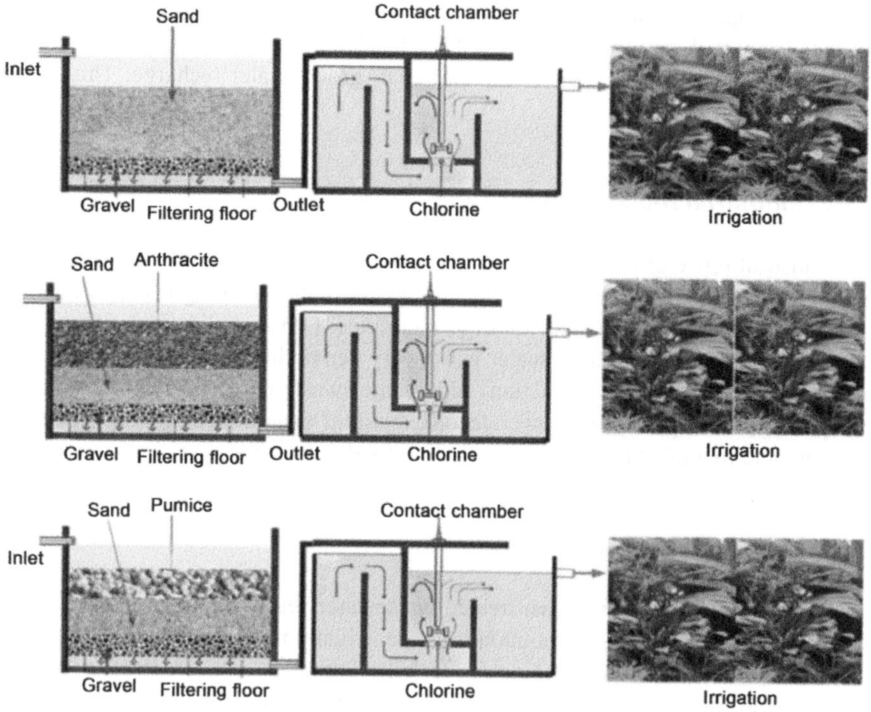

Figure 1.1. *Tertiary treatment with various filters and chlorine disinfection for irrigation purposes*

Non-drinking treated wastewater is used as a new water source to promote agricultural and aquaculture production, industrial uses, reclamation uses such as irrigation, washing and flushing, recreation and artificial recharge. The reuse of effluent from tertiary wastewater treatment increases water sustainability in an environmentally friendly way.

Tertiary treatment is essential for compliance with the World Health Organization (WHO) guidelines and European and national wastewater discharge standards. In other words, the removal of phosphorus and other pollutants from wastewater is a legal obligation. In France, treated wastewater is most often reused for agricultural irrigation and watering green spaces. The efficiency of tertiary treatment is therefore essential to guarantee water quality prior to reuse. The various methods used ensure the complete removal of undesirable substances, allowing predefined quality objectives to be met.

For example, if wastewater is to be reused for vegetable irrigation, pathogens must be removed. In urban reuse, on the other hand, nitrogen and phosphorus are targeted during treatment. The same applies to groundwater recharge. This avoids any risk of eutrophication.

1.1.2. *Industrial reuse and recycling*

Industrial reuse of treated wastewater and internal recycling are now a technical and economic reality. Thermal and nuclear power plants (cooling water) are among the sectors that use wastewater in large quantities. Other applications are possible when the quality of the reused water is not very demanding, such as the mechanical and metal industries, construction (aggregate washing, concrete manufacture, equipment cleaning, cooling tower feed, excluding evaporative cooling), chimney cleaning and manufacturing water (excluding food processing).

1.1.3. *Urban non-drinking water*

In urban and suburban areas, treated wastewater reuse can be highly beneficial for municipalities. The most common uses are irrigation of green spaces (parks, golf courses, sports fields), landscaping (waterfalls, fountains, ponds), street or vehicle washing and fire protection, amenity ponds, swimming pools, ponds for fishing and boating, etc.

Figure 1.2. *Reusing water for non-drinking urban use*

1.1.4. *Storage*

All wastewater reuse plants require a certain amount of storage, in order to regulate daily variations in production flow. In certain sensitive cases, storage is needed to cope with the risk of interrupted production of treated wastewater or failures in the treatment plant. In water-deficient regions in particular, storage is seasonal: it will store water that is not needed in winter, to be used as and when required. It will even out daily variations in the flow from the treatment plant and store the surplus when the average wastewater flow exceeds irrigation requirements.

Storage is therefore an essential link between the treatment plant and the irrigation system. It is also used to provide insurance against the possibility of inappropriate wastewater entering the irrigation system.

1.2. Nosedo plant (Milan, Italy): irrigation

The Nosedo wastewater treatment plant (Milan, Italy) treats the wastewater of 1,250,000 population equivalent (P.E.) with nitrification and denitrification. It is the largest plant in Europe, with almost 60–70% of treated wastewater destined for agricultural use:

– 432,000 m^3 treated per day;

– 5 m^3/s treated in dry weather;

– 15 m^3/s in rainy weather;

– 60–70% of treated water is used for agricultural purposes.

Urban wastewater comes from the central-eastern area of Milan, from the discharger emissary of Nosedo and the Ampliamento Est collector.

The Milano-Nosedo wastewater treatment plant is one of Europe's largest. It treats 50% of Milan's wastewater, in particular from the central and eastern parts of the city. The plant is capable of treating around 157,000,000 m^3/year of wastewater, which is then returned to the water system (Roggia Vettabbia), but also reused for irrigation purposes.

During the design phase of this plant, great attention was paid to certain aspects related to energy efficiency, such as the choice of equipment with inverters, the installation of a high-efficiency fine-bubble aeration system for oxygen transfer, and O_2 probes to monitor and control oxygen concentration in each biological treatment line. It should also be noted that the Nosedo plant features biological oxidation/

nitrification stages with a high sludge age, tertiary sand filtration and high-level disinfection.

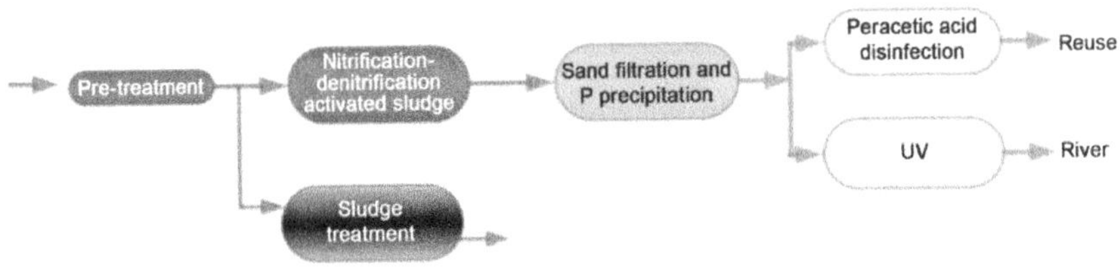

Figure 1.3. *Wastewater treatment plant with reuse (Nosedo, Milan)*

The plant was commissioned on October 30, 2004.

Treated wastewater is used to irrigate rural areas and is accepted as a valuable resource by farmers.

The wastewater treatment plant is an excellent example of how to reduce the demand for water from conventional sources and make efficient use of water resources in many areas. Indeed, the project also includes improvements to the surrounding areas aimed at recomposing the landscape and environmental features of part of the historic Vettabbia valley, dominated by the Chiaravalle Abbey, and creating diverse habitats that are interconnected to maximize biodiversity.

Figure 1.4. *Nosedo wastewater treatment plant including tertiary treatment*

During the irrigation period (April–September), water is discharged into a ditch called Roggia Vettabbia Bassa. This water is then diverted to the fields through channels.

In the area served by the Emissario Nosedo, Roggia Vettabbia and Redefossi agricultural water networks, the farming system is mainly extensive and traditional, with forage crops (grass, corn), rice and cereals.

Parameters	Raw water inlet ($mg.L^{-1}$)	Treated water outlet ($mg.L^{-1}$)	% of reduction
BOD5	170	<5	99
COD	300	<15	97
Total N	27	6.5	76
Total P	3.5	0.9	74
TSS	190	<5	99
Microorganisms (CFU per 100 mL)			
Total coliforms	10^5	40	–
E. coli	10^4	<10	–

Table 1.3. *Water quality at the outlet of the Nosedo tertiary treatment plant*

Water quality parameters for influent and effluent are shown in Table 1.3.

Figure 1.5. *On-site spraying with reused water (Nosedo)*

Operating costs for the Nosedo wastewater treatment plant amount to €0.139/m³. The main component of operating costs is the cost of chemicals (30%), including peracetic acid used for disinfection. Other operating costs include labor (20%), energy (20%), sludge disposal (15%) and maintenance (15%).

The main feature of the Nosedo plant is the large volume of treated water that is reused for agricultural purposes.

The Nosedo wastewater treatment plant is one of the best examples of water reuse in Europe. The challenge was to guarantee the best conditions for water reuse. This is why the treated wastewater complies with European directives.

The plant is equipped with software for real-time optimization of process performance. It provides control data to optimize the entire wastewater system, including the sewer network and treatment plant, energy consumption, chemicals, as well as biological and hydraulic capacity improvement.

The farmers' associations known as Consorzio Vettabbia comprise 84 member farms.

1.3. The Hermitage station (Reunion Island)

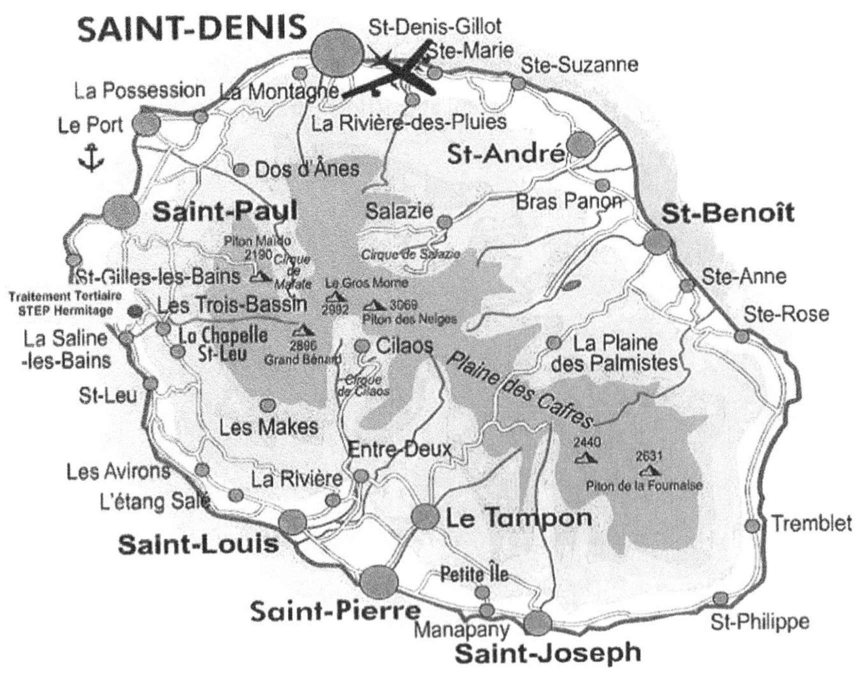

Figure 1.6. *Map of Reunion Island (France)*

Improving local management of water resources is one of Réunion's fundamental challenges in terms of sustainable development.

The targets set are still based on supplying the population with adequate quality drinking water, as well as on efficient wastewater treatment and reuse of treated water for irrigation or industrial purposes.

For example, the rehabilitation program for the Hermitage wastewater treatment plant involved bringing the plant's discharges up to standard, and doubling its treatment capacity from 12,500 to 25,000 P.E. in 2010.

1.3.1. *Pollutant loads*

The basic pollutant load data taken into account for tertiary treatment are defined by hydraulic load parameters.

1.3.1.1. *Hydraulic load*

The maximum hydraulic flow rate is 300 m^3/h for:

– 6,500 m^3/day for 22 h in dry weather;

– 14,000 m^3 for 47 h in wet weather.

1.3.1.2. *Treatment targets*

The system is designed to meet the requirements for water intended for irrigation and potentially for non-food industrial use.

The tertiary treatment process includes the following facilities:

– a lift station;

– a technical building with:

  - electrical room, reagent room, air and backwash filter equipment,

  - wash water tank,

  - sand filter and UV reactor,

  - venturi meter with discharge to existing drain,

  - a wastewater discharge station.

The tertiary treatment lift station is fed by gravity from refining lagoon No. 3. A 3-mm mesh strainer is placed on the station's feed pipe to stop large particles and miscellaneous debris. This strainer is self-cleaning, thanks to a ramp equipped with nozzles for countercurrent injection of pressurized water. Pressurized water is drawn from the pumping station's discharge manifold through a "tapped" pipe.

The self-cleaning strainer is cleaned by a pump connected directly to the storage tank. Four pumps (3 + 1 back-up) supply the tertiary treatment system in 100 m^3/h flow steps. The start-up of one, two or three pumps is controlled by a level measurement on the infiltration lagoon.

The nominal flow rate of 300 m^3/h is discharged through two in-line, self-cleaning pre-filters with a mesh size of 50 µm, each rated at 150 m^3/h. Turbidity is measured downstream of the pre-filters, and a coagulant, such as ferric chloride, is injected at a configurable turbidity threshold.

Raw water and coagulant are mixed by an in-line static mixer. The pressurized sand filters are fed from a general feeder with pre-filtered and possibly pre-coagulated water.

1.3.2. *Tertiary filtration*

An in-line injection of ferric chloride upstream of the sand filters is therefore projected. Coagulation is triggered when the total suspended solids (TSS) concentration of the water pumped into lagoon No. 3 exceeds 25 $mg.L^{-1}$. Injection of ferric chloride is triggered by a turbidity threshold in the water pumped into lagoon No. 3. In addition to reducing particle-bound phosphorus, ferric chloride injection also slightly reduces dissolved phosphorus precipitated by the iron salt. Ferric chloride is delivered in 1 m^3 containers.

The coagulant is injected by a dosing pump (1 + 1 backup) into the main line, upstream of the static mixer. The static mixer, made up of helicoids, ensures turbulence for homogenization and good coagulation. Reagent injection is not continuous and depends on the quality of the water fed to the tertiary filters.

Tertiary filtration is carried out using six pressurized filters, fed by the water lift station of lagoon No. 3, after pre-filtration and possibly pre-coagulation. The filter layer is made up of 1.35 mm expanded clay media (Biodagene) resting on a strained floor.

1.3.2.1. *Choice of filtration*

The choice of particle size – 1.35 mm – and filter bed height – 2 m – offers the following advantages:

– prevents rapid clogging of the surface filter, by favoring a higher layer height for TSS trapping;

– prevents significantly reduced filtration cycles in the event of TSS peaks, and limits degradation of the filtered water;

– prevents filtration speeds of 15 $m.h^{-1}$.

Stable filtered water quality is also beneficial for UV disinfection.

The water to be treated is fed into the filters through a gully at the top, which has a dual role: intake of the water to be filtered and discharge of the wash water.

The filter media consists of a 2-m thick layer of Biodagene (Table 1.4). Beneath the filter layer is a 10-cm layer of gravel and a metal floor fitted with nozzles.

The floor ensures perfect homogeneity of fluid distribution within the filter mass, both during filtration itself and during washing operations.

The filtered water is discharged through a pipe at the bottom of the filter. This pipe also serves as a water inlet for filter washing. The design of the system allows simultaneous washing and filtration. Filters are washed in turn during or outside operating periods.

Washing is triggered by a timetable, clogging threshold or operator decision. Filter pressure loss is checked by measuring the difference in pressure between the filtered water inlet pipe and the filtered water outlet pipe (contact differential pressure gauge).

Filters are preferably washed at night, when the flow treated at the WWTP is reduced. This will limit the hydraulic impact on the plant.

1.3.2.2. *Filter cleaning*

The various washing operations take place automatically. A wash cycle comprises several distinct phases:

– First phase: detasseling. Air is blown in at a rate of 50 $Nm^3.h^{-1}.m^{-2}$ for 1 min.

– Second phase: air + water. Air is blown in at a rate of 50 $Nm^3.h^{-1}.m^{-2}$ combined with a countercurrent of water limited to 8 $m^3.h^{-1}.m^{-2}$ for 8 min.

– Third phase: rinsing and evacuation of dirty water with water alone at a flow rate of 50 $m^3.h^{-1}.m^{-2}$ for 8 min.

Wash water is run through the same pipe as the filtered water outlet. Air is supplied through a pipe running under the floor. Wash water is supplied by a pump providing the required flow rate of 200 $m^3.h^{-1}$ for maximum manometric head.

The wash water is pumped into the 20 m^3 filtered water tank and disinfected. Dirty water is then collected in a station equipped with two 60 $m^3.h^{-1}$ unitary drainage pumps (1 + 1 backup) discharging to the biological tank.

Parameters	Unit	Values
Total incoming flow (over 15 h)	$m^3.h^{-1}$	300
Number of filters	u	6
Filtration rate on N filters	$m.h^{-1}$	13.2
Filtration rate on N^{-1}	$m.h^{-1}$	15.8
Unit surface area	m^2	3.8
Unit diameter	m	2.2
ES Biodagene	mm	1.35
Height of material (gravel + Biodagene)	m	0.1 + 2.0
Minimum cycle time (TSS input 25 $mg.L^{-1}$)	h	24

Table 1.4. *Tertiary filtration dimensions*

1.3.3. *UV disinfection*

Figure 1.7. *UV reactors and filters*

The principle of disinfection by UV radiation has developed considerably over the last few decades for wastewater treatment, and it is considered a good alternative to chlorination. With UV radiation, disinfection is not accompanied by the formation of any reaction products with organic matter in the water. UV rays are light waves with wavelengths between 100 and 400 nm. Their germicidal power depends on the wavelength emitted. UVC between 200 and 280 nm is the most germicidal.

Low-pressure lamp technology was chosen for this installation. These have the advantage of emitting maximum radiation in the wavelength corresponding to the optimum germicidal band (254 nm):

– long service life: 15,000 operating hours (12,000 guaranteed operating hours), 15,000 reported hours;

– low-power consumption;

– 96% of the spectrum is emitted at 253.7 nm for maximum efficiency;

– the applied dose is 400 $J.m^{-2}$ at the end of the lamp life.

Three lines (Figure 1.7), each equipped with a 100 $m^3.h^{-1}$ in-line low-pressure UV reactor, are installed downstream of the tertiary filtration unit. They are independent and can be isolated from one another. The system is controlled by an electrical cabinet, which switches on the lamps, operates them and counts up the operating hours, and is equipped with an alarm to indicate any malfunction.

The reactors (Table 1.5) feature automatic lamp cleaning by sweeping. The cleaning frequency is adjustable according to water quality and intensity measurement.

The UV reactors are as follows:

– arranged horizontally to facilitate inspection and maintenance operations;

– equipped with two draining and cleaning valves;

– equipped with three mixers to homogenize effluent and disinfection.

In addition, turbidity is measured to assess the presence of suspended matter in the filtered water, which is known to reduce UV efficiency. The UV disinfection stage can be bypassed by a set of manual valves in the event of significant degradation of the filtered water.

Treated and disinfected water is collected in a buried tank, and then pumped through a venturi channel. A flow-controlled refrigerated sampler is provided.

Parameters	Values
Flow rate	100 $m^3.h^{-1}$
Installed germicidal power	1,500 UVC watts
Water volume in chamber	107 L
Number of lamps per reactor	10 lamps
Average lamp life	15,000 h
Contact time	3.85 s
Installed capacity per reactor	3.7 W
Approximate pressure loss	0.06 bar
Number of UV reactors	3

Table 1.5. *UV reactor dimensions*

The results are as follows:

– The TSS concentration in produced water was always below 5 $mg.L^{-1}$, and most often around 2 $mg.L^{-1}$.

– The results obtained on the produced water, with regard to microorganisms, show that the concentrations obtained with the various microorganisms are well below the required thresholds in the majority of cases. Interestingly, the values for salmonella and enterovirus are zero. Helminth egg values were always below 1.L^{-1}.

Concentrations of 0–1 CFU/100 mL were obtained with fecal streptococci, with a high value of 6 CFU/100 mL.

The expected bacteriological quality after UV treatment is shown in Table 1.6.

At certain short periods, higher suspended solids values in the treated water (>5 $mg.L^{-1}$) were observed, resulting in a slight deterioration in disinfection conditions. However, concentrations of fecal coliforms and streptococci remained below guarantee levels.

The UV transmittance value of the water ranged, most of the time, between 58% and 62%, indicating low suspended solids and low water coloration. These parameters justify the efficiency of filtration and their implications for UV disinfection performance.

Parameters	Water quality at wastewater treatment plant outlet	Values obtained at tertiary treatment outlet
BOD5	25 mg.L^{-1}	–
COD	125 mg.L^{-1}	–
TSS	35 mg.L^{-1}	<5 mg.L^{-1}
Total N	15 mg.L^{-1}	–
Total P	2 mg.L^{-1}	–
Turbidity	10 NTU	<3 NTU
Total coliforms (CFU/100 mL)	10^7	0–50
Fecal coliforms (CFU/100 mL)	10^6	0–20
Fecal streptococci (CFU/100 mL)	10^4	100
Salmonella (CFU/100 mL)	10^3	0–1
Enteroviruses (per L)	10^4	0 per 10 L
Intestinal helminth eggs (per L) ***Ascaris*** **and** ***Trichuris*** **species**	–	<1

Table 1.6. *Water quality at tertiary treatment inlet and outlet*

Figure 1.8. *Tertiary treatment plant (Reunion Island, France)*

1.3.4. *Educational aspect*

All tertiary treatment pipes have been painted with a very precise color code. The aim is to use this installation as an educational tool. The tertiary treatment plant is frequently visited by schools, colleges, universities, associations and teachers. Color-coded piping makes it easy for all visitors to understand.

This example of tertiary wastewater treatment at the Hermitage plant (Réunion) shows that it is one of the possible solutions for reusing this water to cope with recurring periods of drought and/or limit water withdrawals from a natural environment. In this way, the water produced can be reused for other purposes (such as irrigation) instead of drinking water resources.

It also makes it possible to remove or defer discharges from the treatment plant into vulnerable environments. With conventional technologies such as sand filtration and UV disinfection, it is possible to respect water quality standards with a view to agricultural and industrial reuse, or to protect recreational areas (bathing waters, sporting activities). This is a major issue at the Hermitage station, which is a major tourist area on the island.

1.4. The Barcelona plant (Spain): regeneration of wetlands, irrigation

The Llobregat River is one of the main sources of water supply for the Barcelona metropolitan area. The Catalan Water Agency requires that a minimum water flow of 3.5 $m^3.s^{-1}$ be maintained in summer. Irregular rainfall in the Llobregat watershed has led to a profound water deficit in the river and in the surrounding areas. As a result, emergency measures were implemented, with seasonal restrictions on irrigation water.

A wastewater reuse project was therefore planned with the aim of:

– replenishing the water resources vital to Spain's Mediterranean coast, particularly in the metropolitan area of Barcelona, home to 4 million inhabitants;

– improving the environmental balance in the Llobregat delta by finding an alternative to discharging 153 million m^3/year of fresh water into the sea;

– solving the growing problem of water shortages in the dry season, while preserving the ecology of the Llobregat delta and river, and protecting groundwater that has been overexploited or degraded by saline intrusion. To achieve this, it was necessary to find an additional 50 million m^3 of water per year.

With a capacity of 302,400 $m^3.day^{-1}$, it is the largest wastewater reuse plant in Europe and one of the largest in the world. This treatment process is designed to:

– increase available water resources: 64,800 $m^3.day^{-1}$ to irrigate agricultural areas;

– restore and recharge 34,560 $m^3.day^{-1}$ of wetlands in the delta;

– return 172,800 $m^3.day^{-1}$ to the Llobregat river to restore environmental balance.

The water recycled in this way maintains the river's flow at an ecologically satisfactory level and meets the water requirements for agricultural irrigation and recharging the delta's wetlands.

The quality of reused water is different for each use. The criteria used for the various applications are shown in Table 1.7.

Wetlands and irrigation	Salt water intrusion barrier
BOD5: ≤10 $mg.L^{-1}$ TSS: ≤5 $mg.L^{-1}$ Turbidity: <5 NTU Fecal coliforms: <10 UFC/100 mL Intestinal helminth eggs: <1 u/100 mL Residual chlorine : >0.6 $mg.L^{-1}$ Dissolved oxygen: ≥7.5 $mg.L^{-1}$	TSS ≤ 1 $mg.L^{-1}$ Turbidity: <0.1 NTU Organic matter: <10 $mg.L^{-1}$ Fecal coliforms: 0 per 100 mL

Table 1.7. *Water quality requirements for wastewater reuse at the Barcelona site*

Depending on the uses of the reused water, two technical solutions are applied at the Catalan wastewater treatment plant.

The first solution consists of a conventional tertiary treatment process comprising coagulation-flocculation-settling (3 Actiflo), followed by mechanical filtration (10 Hydrotech disk filters) and ultraviolet disinfection, and lastly, post-disinfection with chlorine.

The second solution uses a double-membrane solution (see Chapter 4) that delivers a more stringent water quality, intended for reinjection into the ground and to restrict salt intrusion.

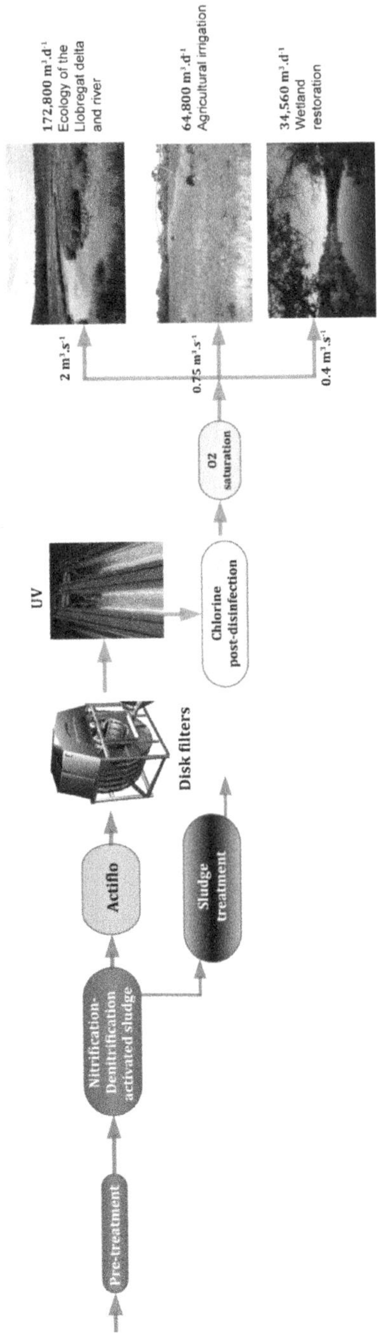

Figure 1.9. *Treatment plant for wastewater reuse in Barcelona*

Figure 1.10. *Barcelona plant with tertiary treatment*

The results obtained show that the quality of treated water intended for various reuse applications meets the expected objectives. Microbiological quality is excellent after the two stages combined: UV disinfection followed by chlorination.

Parameters	Units	Values
TSS	$mg.L^{-1}$	3.0–5.0
Turbidity	NTU	1.0–1.6
BOD5	$mg.L^{-1}$	< 3
COD	$mg.L^{-1}$	22–44
N-NH$_4$	$mg.L^{-1}$	0.9–2.5
N-NO$_3$	$mg.L^{-1}$	4.2–6.9
Total N	$mg.L^{-1}$	6.9–10.4
Total P	$mg.L^{-1}$	0.6–1.5
Escherichia coli	CFU.100 mL^{-1}	<1
Helminth eggs	u.10 L^{-1}	<1
Fecal coliforms	CFU.100 mL^{-1}	<1
***Legionella* spp.**	$CFU.L^{-1}$	<100

Table 1.8. *Performance of the Barcelona wastewater reuse plant*

In the reused effluent, a variety of contaminants, such as pharmaceuticals and personal care products (PPCPs) and disruptive compounds (EDCs), have been detected in a wide range of concentrations. A quaternary RO membrane treatment ensures their removal, with the produced water reinjected into the subsoil.

1.5. Ajman (United Arab Emirates): irrigation and non-drinkable urban use

The United Arab Emirates has one of the highest water consumption rates in the world, with each person using an average of 550 L of water per day. Total annual water demand is expected to reach 8.8 billion cubic meters by 2030, almost double what is consumed today.

As an alternative to the consumption of non-renewable sources, a comprehensive wastewater reuse plan has been drawn up for use in residential and recreational areas, green spaces and for industrial purposes. The wastewater treatment plant has a capacity of 140,000 $m^3.day^{-1}$. As a result, one-third of the treated wastewater is reused to irrigate various areas of the Emirate of Ajman or for industrial applications.

The authorities' aim is to make reused water a norm throughout the Emirate of Ajman, and for the general public to accept it as one of the best available sources for sustainable water use. The general idea is that the people of Ajman and the United Arab Emirates will be more open to the idea that treated wastewater can be reused.

Figure 1.11. *Wastewater reuse in Ajman (United Arab Emirates)*

The Ajman wastewater reuse plant is based on the double-pass membrane principle, comprising a first microfiltration stage followed by a second RO stage. The water produced is used for various industrial and commercial purposes 24/7.

Distribution takes place via a tank-filling area that can simultaneously accommodate 12 tanks of different capacities based on the service-station model.

Water quality parameters such as TDS (total dissolved solids) are better than most commercial bottled waters, making it perfect for demanding industrial use.

Figure 1.12. *Reverse osmosis membrane and service station for reused wastewater*

Figure 1.13. *Distribution of treated wastewater by "non-drinking water" tanker trucks*

Once treated, the wastewater is reused for unrestricted irrigation of Ajman's entire landscape and greenery. As the plant produces high-quality water, it has been adopted by many industrial companies for non-drinking uses, as a substitute for groundwater. Customers include air-conditioning plants, civil engineering companies and non-food industries, all of which have the common requirement of very low salinity water.

1.6. Al Wathba treatment plant (Abu Dhabi): recreational areas, green spaces, industries

The Emirate of Abu Dhabi is an arid region with an average annual rainfall of less than 100 mm. The Emirates' water resources are traditional or conventional (rainfall, springs, wadis, sabkhas, lakes, ponds and groundwater) and non-traditional or non-conventional (desalinated water and treated wastewater).

Groundwater supplies are dwindling, and the imbalance between supply and demand is resulting in ever-increasing quantities of desalinated water.

Water shortages for irrigation are offset by desalinated water. Despite investments in desalination, this has had no impact on the desire to save drinking water, as it is free for municipal use.

This proved crucial for the United Arab Emirates, where groundwater resources are limited and the urban population is growing rapidly. The combination of high population growth and rapid urbanization meant that the United Arab Emirates had to meet two major challenges: to treat the growing quantity of wastewater produced and provide ever more water resources for its population, industry and agriculture. As part of its *Abu Dhabi Economic Vision 2030* development plan, the Emirates has made environmental management one of its top priorities. The treatment and reuse of wastewater has become a cornerstone of the Emirates' sustainable development.

In Abu Dhabi, to complement existing wastewater treatment facilities, large-scale plants have been built in joint venture with the private sector (BOOT mode). These include Al Wathba, Allahamah and Al Saad, which are designed to supply water for irrigation.

Tertiary treatment is the final wastewater treatment process at the Al Wathba 2 wastewater treatment plant. The tertiary treatment process comprises two stages: filtration and disinfection.

A two-layer filter (pumice and sand) is used to remove suspended particles larger than 10 μm. The filtered effluent is then disinfected by sodium hypochlorite injection. Sodium hypochlorite is generated on-site by electro-chlorination units.

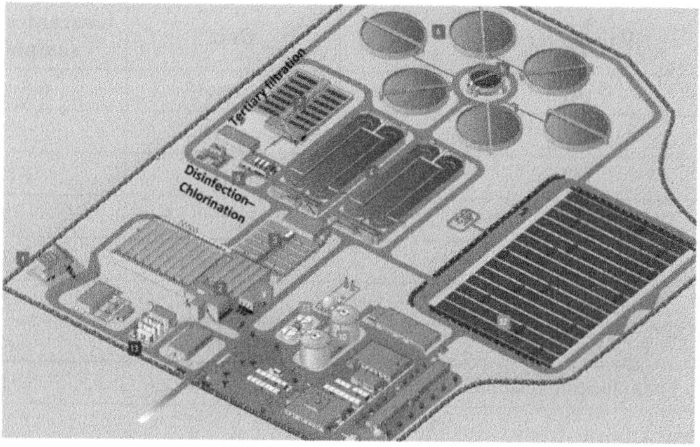

Figure 1.14. *Al Wathba wastewater treatment and reuse plant*

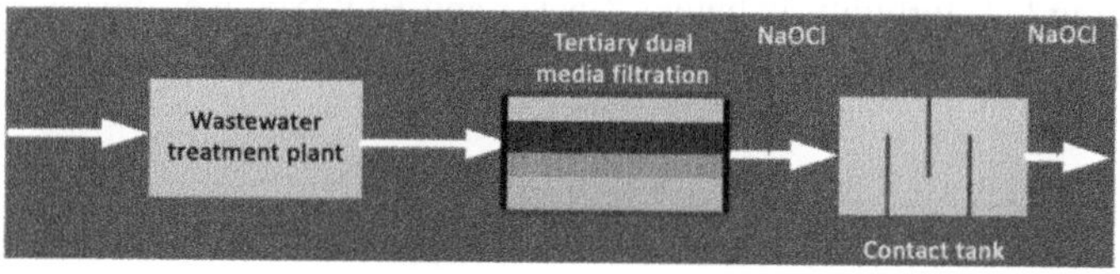

Figure 1.15. *Tertiary treatment plant (Al Wathba, Abu Dhabi)*

Parameters	Units	Values
Daily flow rate (contractual capacity)	$m^3.day^{-1}$	300,000
Hourly flow rate	$m^3.h^{-1}$	12,500
Fecal coliforms tertiary treatment inlet	CFU/100 mL	$\leq 10^7$
Contact tank volume	m^3	6,459
Average contact time	min	31
Minimum contact time	min	24.2
Chlorine consumption	$kg.h^{-1}$	203.8 Max.: 348.8

Table 1.9. *Operating conditions for tertiary treatment*

The disinfection step is based on the following operating conditions presented in Table 1.9.

Parameters	Unit	Guarantee (average sample 24 h)
pH	–	6.5–8.5
BOD5	mg.L^{-1}	≤10
TSS	mg.L^{-1}	≤10
N-NH4	mg.L^{-1}	≤2
Dissolved oxygen	mg.L^{-1}	≥3
Turbidity	NTU	≤2
E. coli	CFU/100 mL	≤10
Free chlorine residual	mg.L^{-1}	≥1

Table 1.10. *Fecal coliform guarantees at the Al Wathba plant (Abu Dhabi)*

The results obtained show that guarantees are respected for all parameters. TSS is below 3 mg.L^{-1}, turbidity is below 2 NTU and BOD5 is well below 10 mg.L^{-1}. *Escherichia coli* concentrations are at 0 CFU per 100 mL more than 99% of the time.

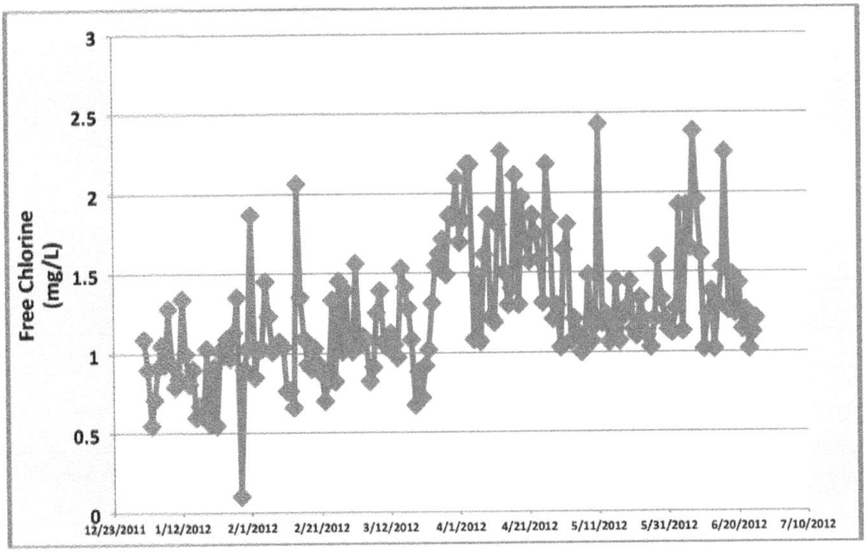

Figure 1.16. *Free chlorine residual at tertiary treatment outlet*

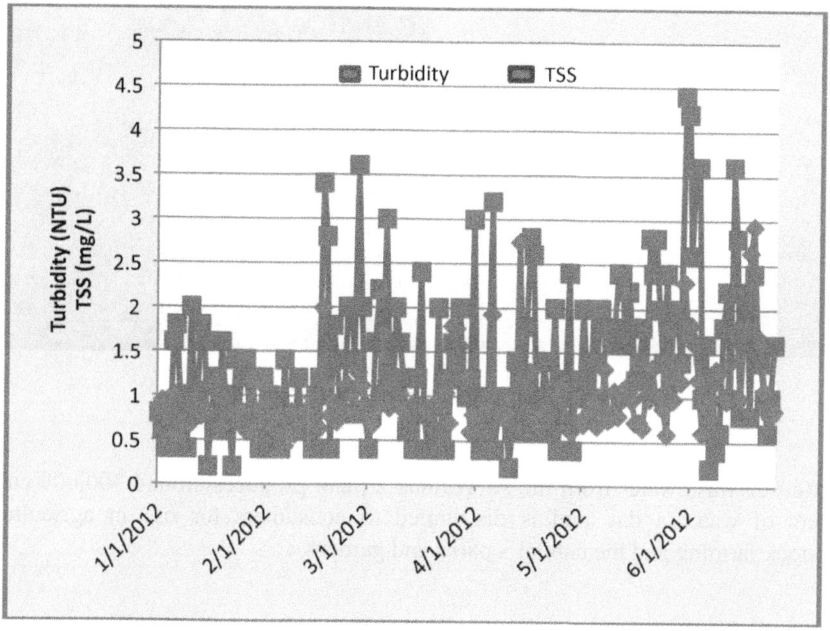

Figure 1.17. *Changes in turbidity and TSS in reused water*

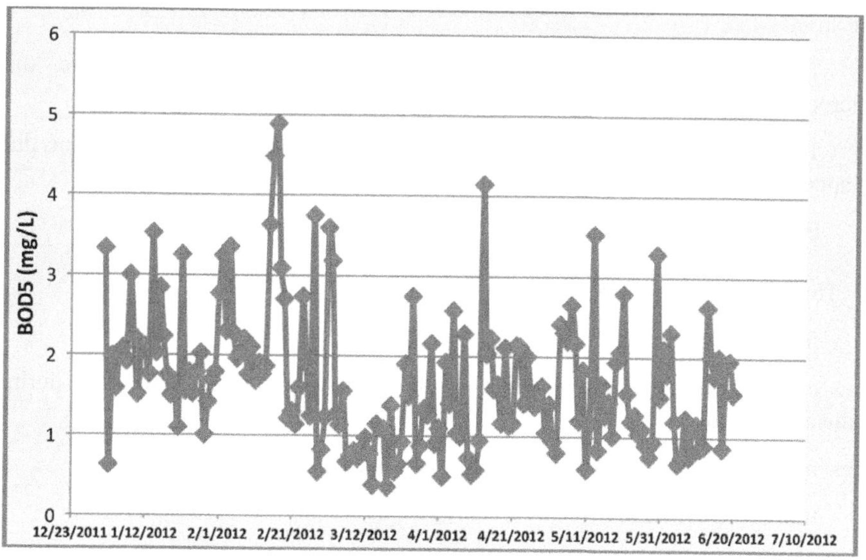

Figure 1.18. *Changes in BOD5 at tertiary treatment outlet*

Figure 1.19. *Applications for reused wastewater*

Reused wastewater from the Al Wathba 2 plant produces around 300,000 cubic meters of water a day and is distributed to consumers for use in agriculture, livestock farming and the capital's parks and gardens.

1.7. Burj Khalifa lake (Dubai): reuse wastewater for a recreational area

The project involves reusing rainwater and concentrate from a demineralization plant to supply the Burj Khalifa lake (Dubai). The requirements of the specifications are strict:

– The treatment system must maintain water clarity and ensure that microorganisms, algae, fungi, etc., do not grow in the lake.

– The water must, at all times, be free of harmful bacteria, pale blue in color, dull in appearance and odorless.

– Biocides and algaecides are permitted where necessary.

The treatment process includes:

– feed water pumping;

– coagulation-flocculation-settling with ballasted flocs (Actiflo®) during rainwater periods;

– mechanical filtration on disk filters (10 µm cut-off);

– transfer to the lake.

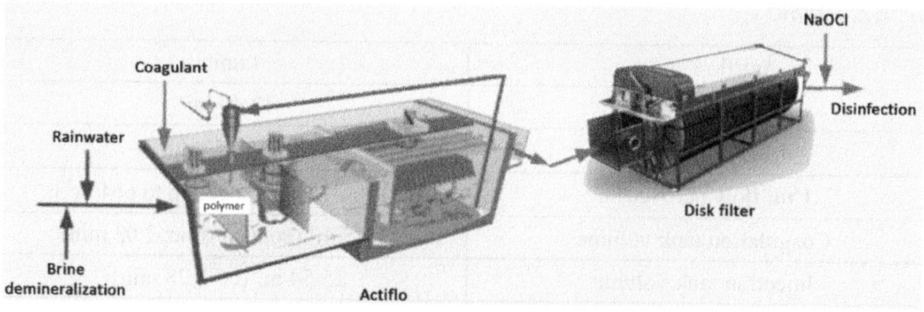

Figure 1.20. *Treatment plant for Burj Khalifa lake (Dubai)*

To prevent dissolved salts from demineralization brines from accumulating in the lake, the Actiflo® does not operate continuously. It only operates during storms, when the water is loaded with suspended solids.

No chemicals such as PACl (polyaluminum chloride) are injected into the water treatment line under dry conditions. PACl is dosed during storm conditions, so as to limit the concentration of salts in the lake.

The injection of chemical reagents during stormy periods is necessary, but does not lead to an increase in salts, as rainwater dilutes the water present in the lake.

When stormwater arrives at the treatment plant, Actiflo® units treat it as a priority. The inlet system can handle a total flow of 2,600 $m^3.h^{-1}$.

1.7.1. *Treatment plant dimensions*

1.7.1.1. *Recreation area (lake)*

Water level in lake	1.3 m
Lake surface area	127,875 m^2
Volume of water in lake	166,238 m^3 (treated in 3 days)
Water flow to be treated in lake	2,309 $m^3.h^{-1}$ = 166,238 m^3/3 days

Table 1.11. *Dimensions of the recreational lake*

1.7.1.2. *Actiflo*

Actiflo® units	4 units
Mirror surface	12.40 m^2
Mirror speed	51.0–53.5 m^3.h^{-1}
Unit flow per Actiflo	630 m^3.h^{-1} on average, up to 664 m^3.h^{-1}
Coagulation tank volume	22,45 m^3 (contact time: 2.03 min)
Injection tank volume	25.54 m^3 (ct: 2.28 min)
Flocculation tank volume	38.70 m^3 (ct: 3.45 min)
Settling tank volume	67.50 m^3 (ct: 6.03 min)

Table 1.12. *Actiflo dimensions (Veolia)*

1.7.1.3. *Mechanical filtration*

Disk filter (Hydrotech)	4 units
Number of trays per unit	14 trays
Unit wash flow	14 m^3.h^{-1} on average, 20 m^3.h^{-1} max. per unit
Total wash volume	$4 \times 14 = 56$ m^3.h^{-1} (on average)
Wash flow rate	14 m^3/h per unit (continuous washing)
Maximum wash flow	20 m^3/h per unit (exceptional washing)

Table 1.13. *Disk filter dimensions (Hydrotech)*

1.7.1.4. *Disinfection*

NaOCl is injected into the water from the demineralization concentrates that feed the reuse treatment unit. NaOCl levels averaged 2.0 mg.L^{-1}, with a maximum of 3 mg.L^{-1}. Actiflo sludge and wash water are collected and sent to the sludge treatment plant.

1.7.1.5. *Reagent dosages*

Polyaluminum chloride (PACl), ferric chloride ($FeCl_3$) or ferric sulfate ($Fe_2(SO_4)_3$) reagents can be used as required.

Note that 207 mg.L^{-1} of Al polychloride at 8.5% and 1.20 density are equivalent to 137 mg.L^{-1} of $FeCl_3$ at 41% and 1.45 density.

A total of 137 mg.L^{-1} of $FeCl_3$ at 41% and 1.45 density are equivalent to 160 mg.L^{-1} of $Fe_2(SO_4)_3$ at 43% and 1.45 density.

1.7.1.6. *Quality of water produced*

TSS (average)	5 mg.L^{-1}
TSS (maximum)	8 mg.L^{-1}
Turbidity	<5 NTU
Color	<20 mg.L^{-1} Pt/Co
Total coliforms	<100 CFU/100 mL

Table 1.14. *Quality of water produced*

Figure 1.21. *Burj Khalifa lake water quality with reused wastewater*

Figure 1.22. *Water features on the lake*

1.8. Darling Quarter (Sydney, Australia): wastewater reuse in a neighborhood

Sydney's newest landmark, Darling Quarter-Commonwealth Bank Place, consists of two new low-rise commercial buildings. It comprises an eight-storey Grade A office space (55,418 m^2), a commercial ground floor (3,789 m^2), 800 parking spaces, as well as a children's playground and multi-purpose theater.

Figure 1.23. *Darling Quarter (Sydney)*

Veolia was selected to design, build and operate a recycled water plant for the Darling Quarter. The project offered a 90% reduction in drinking water thanks to rainwater harvesting and on-site wastewater recycling.

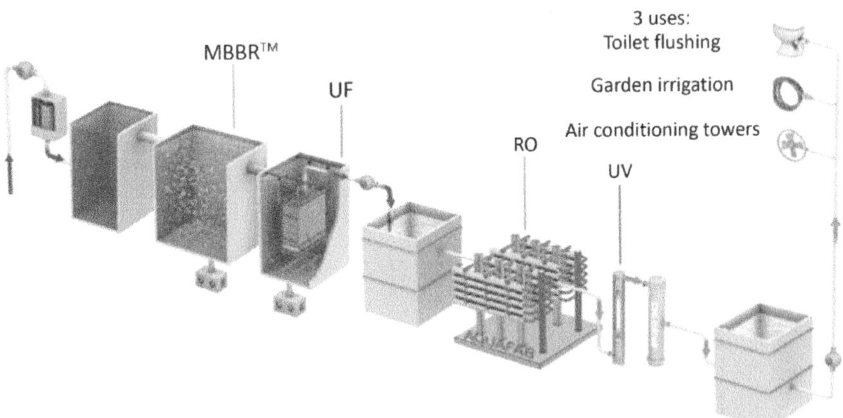

Figure 1.24. *Wastewater reuse in a neighborhood*

The treatment process combines the following stages involving innovative technologies:

– moving bed biofilm reactor (MBBR);

– Neosep® submerged MBR;

– RO unit;

– disinfection.

Figure 1.25. *Reuse station equipment*

Effluent is extracted into a sewer pipe running underground on the site, which diverts wastewater from the main pipe to a secondary pipe, using a weir bypass system. This solution was preferred to on-site pumping, as it allows for smaller storage and better control of the intake, so that if the treatment plant is offline, the intake is stopped.

The process treats 245 m^3 of wastewater per day through sewer extraction, producing 166 m^3 per day (60,000 m^3/year) of high-quality recycled water for reuse in flushing, irrigation and cooling tower make-up water. From this, 116 m^3 per day is for cooling towers and 50 $m^3.day^{-1}$ is for flushing and irrigation. However, these flows may vary according to the season.

MBBR and Neosep® MBR technologies are ideal for wastewater treatment to withstand load variations and can provide high removal of bacteria, viruses and suspended solids. RO removes salts from feed water. The recycled water plant has been designed to adapt to the spatial constraints of the site.

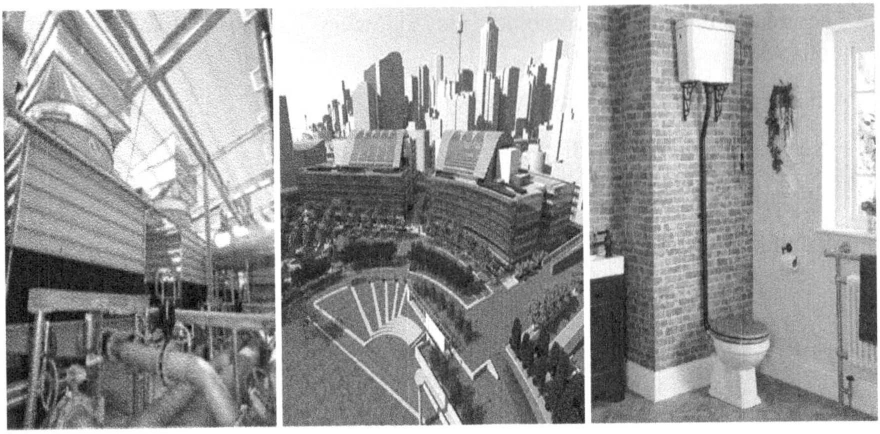

Figure 1.26. *Reuse for cooling towers, irrigation and toilet flushing*

Most of the BOD5, COD and microbiological constituents are greatly reduced in the MBR. The remaining constituents are then removed by RO membranes. Final disinfection ensures microbiological safety.

The quality of the water produced is well above Australian guidelines for wastewater reuse.

Recycled water quality is monitored 24/7, 365 days a year by critical control points and a remote monitoring system. Regular sampling is carried out, and online analyzers ensure that the reused water meets the required quality.

Wastewater reuse, as a sustainable and reliable concept, is a promising solution to alleviate water shortages.

Membrane-based treatment systems are selected on the basis of their hydraulic filtration performance, their capacity to remove various organic and inorganic substances, including micropollutants, their resistance to clogging and their energy consumption. Membrane processes, either alone or in conjunction with other possible processes (activated carbon and advanced oxidation), can be used to further guarantee the quality of the water produced.

The challenges still facing membrane technologies for drinking water applications include institutional constraints, funding and public acceptance.

Parameters	Produced water quality at Darling Quarter	NSW guidelines for private use of wastewater reuse	Australian guidelines for wastewater reuse
BOD5	<5 $mg.L^{-1}$	<10 $mg.L^{-1}$	–
TSS	<5 $mg.L^{-1}$	<10 $mg.L^{-1}$	–
pH	6.5–8.5	6.5–8.5	–
Turbidity	<0.2 NTU	<2 (95%ile), 5 NTU	–
Escherichia coli	<1 CFU/100 mL	<1 CFU/100 mL	<1 CFU/100 mL
Residual chlorine	0.2–2 $mg.L^{-1}$	0.2 $mg.L^{-1}$	–
Clostridium perfringens	<1 CFU/100 mL	<1 CFU/100 mL	–
Coliphages	<1 CFU/100 mL	<1 CFU/100 mL	–
Virus removal throughout the process	6.6 log	–	6.5 log
Bacteria removal for the entire process	12.1 log	–	5 log
Protozoa removal throughout the entire process	8.1 log	–	5 log

Table 1.15. *Comparison of parameters to be removed at the Darling Quarter plant*

1.9. References

Membrane bioreactors

Al-Asheh, S., Bagheri, M., Aidan, A. (2021). Membrane bioreactor for wastewater treatment: A review. *Case Stud. Chem. Environ. Eng.*, 4, 100–109.

Cote, P., Alam, Z., Penny, J. (2012). Hollow fiber membrane life in membrane bioreactors (MBR). *Desalination*, 288, 145–151.

Geenens, D., Thoeye, C., Parmentier, G., Van de Peer, T. (2001). Membrane bioreactors in Flemish wastewater treatment: applications in renovation and reuse. In *Proceedings of AWWA Safe Drinking Water Membrane Technology Conference*. Future of Pure Water, San Antonio.

Hai, F., Yamamoto, K., Lee, C.H. (2014). *Membrane Biological Reactors: Theory, Modeling, Design, Management and Applications to Wastewater Reuse*. IWA Publishing, London.

He, C., Liu, Z., Wu, J., Pan, X., Fang, Z., Li, J., Bryan, B.A. (2021). Future global urban water scarcity and potential solutions. *Nat. Commun.*, 12, 4667–4680.

Hoinkis, J., Deowan, S.A., Panten, V., Figoli, A., Huang, R.R., Drioli, E. (2012). Membrane bioreactor (MBR) technology – A promising approach for water reuse. *Procedia Eng.*, 33, 234–241.

Iorhemen, O.T., Hamza, R.A., Tay, J.H. (2016). Membrane bioreactor (MBR) technology for wastewater treatment and reclamation: Membrane fouling. *Membranes*, 6, 33–45.

Jhansi, S.C. and Mishra, S.K. (2013). Wastewater treatment and reuse: Sustainability options. *Consilience*, 10, 1–15.

Judd, S. (ed.) (2006). The MBR book. Principles and applications of membrane bioreactors. In *Water and Wastewater Treatment*. Elsevier, Amsterdam.

Ladewig, B. and Al-Shaeli, M.N.Z. (2017). *Fundamentals of Membrane Bioreactors*. Springer, Singapore.

Lesjean, B. and Huisjes, E.H. (2008). Survey of European MBR market, trends and perspectives. *Desalination*, 231, 71–81.

Lesjean, B., Rosenberger, S., Schrotter, J.C. (2004). Membrane-aided biological wastewater treatment – An overview of applied systems. *Membr. Technol.*, 56, 5–10.

Pinnekamp, J. and Friedrich, H. (2006). Membrane technology for waste water treatment. *FiW-Verlag*, 1, 6–17.

Radjenovic, J., Matosic, M., Mijatovic, I., Petrovic, M., Barceló, D. (2014). Membrane bioreactor (MBR) as an advanced wastewater treatment technology. In *Emerging Contaminants from Industrial and Municipal Waste*, Barceló, D. and Petrovic, M. (eds). Springer, Berlin/Heidelberg.

Rahman, T.U., Roy, H., Islam, M.R., Tahmid, M., Fariha, A., Mazumder, A., Tasnim, N., Pervez, M.N., Cai, Y., Naddeo, V. (2023). The advancement in membrane bioreactor (MBR) technology toward sustainable industrial wastewater management. *Membranes*, 13, 181–193.

Skoczko, I., Puzowski, P., Szatyłowicz, E. (2020). Experience from the implementation and operation of the biological membrane reactor (MBR) at the modernized wastewater treatment plant in Wydminy. *Water*, 12, 3410–3416.

Sutton, P. (2003). Membrane bioreactors for industrial wastewater treatment: The state-of-the-art based on full scale commercial applications. *Proc. Water Environ. Fed.*, 23–32.

Welo, U., Siagian, R., Khoiruddin, K., Ting, Y.P., Boopathy, R., Wenten, I.G. (2022). Advances in membrane bioreactor: High performance and antifouling configurations. *Curr. Pollut. Rep.*, 8, 98–112.

Xiao, K., Liang, S., Wang, X., Chen, C., Huang, X. (2019). Current state and challenges of full-scale membrane bioreactor applications: A critical review. *Bioresour. Technol.*, 271, 473–481.

Yang, W., Cicek, N., Ilg, J. (2006). State-of-the-art of membrane bioreactors: Worldline research and commercial applications in North America. *J. Membrane Sci.*, 270, 201–211.

Tertiary filtration

Adin, A., Baumann, E.R., Cleasby, J.L. (1979). The application of filtration theory to pilot-plant design. *Jour. Amer. Water Works Assoc.*, 71(1), 17–27.

Amirtharajah, A. (1977). Optimum backwashing of sand filters. *J. Environ. Eng. Div. ASCE*, 104(EE5), 917–938.

Amirtharajah, A. and Cleasby, J.L. (1972). Predicting expansion of filters during backwash. *J. Amer. Water Works Assoc*, 64(1), 52–59.

Armon, R., Mandelbaum, R., Manor, Y., Campos, C., Gillerman, L., Salgot, M., Gerba, C., Klein, I. (2001). Secondary wastewater disposal for crop irrigation with minimal risks. *Water Sci. Technol.*, 43(10), 139–146.

Asano, T. and Cotruvo, J.A. (2004). Groundwater recharge with reclaimed municipal wastewater: Health and regulatory considerations. *Water Res.*, 38, 1941–1951.

Camp, T.R., Graber S.D., Conklin, G.F. (1971). Backwashing of granular water filters. *J. Environ. Eng. Div. Proc. ASCE*, 97(EE6), 903–919.

Carman, P.C. (1937). Fluid flow through granular beds. *Chemical Engineering Research & Design*, 75, S32–S48.

Carman, P.C. (1947). Slip flow in granular beds. *Nature*, 160(4061), 301–302.

Cho, A.G., Dodds, J., Santamarina, J.C. (2006). Particle shape effects on packing density, stiffness and strength – Natural and crushed sands. *Journal of Geotechnical and Geoenvironmental Engineering*, 132(5), 591–602.

Cleasby, J.L. (1993). Declining rate filtration. *Water Science and Technology*, 27(7/8), 11–22.

Crittenden, J.C., Trussell, R.R., Hand, D.W., Howe, K.J., Tchobanoglous, G. (2012). *MWH's Water Treatment: Principles and Design*. John Wiley & Sons Inc., New York.

Culp, R.L. (1977). Direct filtration. *J. Amer. Water Works Assoc.*, 69, 375–391.

Dawoud, M.A., Sallam, O.M., Mahmoud, A., Abdelfattah, M.A. (2010). Treated wastewater management and reuse in arid regions: Abu Dhabi case study water resources department. In *10th Gulf Water Conference*. Environment Agency, Abu Dhabi.

Ergun, S. (1952a). Fluid flow through packed columns. *Chemical Engineering Progress*, 48(2), 78–89.

Ergun, S. (1952b). Determination of geometric surface area of crushed porous solids. *Analytical Chemistry*, 24, 369–388.

Ergun, S. and Orning, A.A. (1949). Fluid flow through randomly packed columns and fluidized beds. *Industrial and Engineering Chemistry*, 41(6), 1179–1184.

Exall, K. (2004). A review of water reuse and recycling, with reference to Canadian practice and potential. *Water Qual. Res. J. Canada*, 39(1), 13–28.

Exall, K., Marsalek, J., Schaefer, K. (2004). A review of water reuse and recycling, with reference to Canadian practice and potential: 1. Incentives and implementation. *Water Qual. Res. J. Canada*, 39(1), 1–12.

Fasciolo, G., Meca, M., Gabriel, E., Morabito, J. (2002). Effects on crops of irrigation with treated municipal wastewaters. *Water Sci. Technol.*, 45(1), 133–138.

Fitzpatrick, C.S.B. (1990). Detachment of deposits by fluid shear during filter backwashing. *Water Supply*, 8, 177–183.

Gaid, K. (2002). Wastewater reuse for the industry. In *Conf. Environment – Industry*. Curitiba University, Brazil.

Gaid, K. and Bonne, P. (2003). Les technologies membranaires dans la réutilisation des eaux usées. *L'eau, l'industrie, les nuisances*, 125, 56–68.

Gaid, K. and Lanning, J. (2000). Wastewater reuse: The case of Durban. In *French-India Conf. on Environment*. Vivendi Water.

Gaid, K. and Le Poder, N. (1998). La réutilisation des eaux usées pour l'agriculture : critères de qualité et moyens techniques. In *IXème Congrès U.A.D.E.*. U.A.D.E., Casablanca.

Gaid, K. and Nugier, L. (2006). Traitement tertiaire des eaux usées en France : le cas de l'usine de l'Hermitage (La Réunion). *L'eau, l'industrie, les nuisances*, 322, 322–342.

Gaid, K. and Zeghal, S. (1998). Wastewater technologies reuse. In *Water & Environment Conference*, Cairo University, Cairo.

Graham, N.J.D. (1988a). Filter pore flocculation as a mechanism in rapid filtration. *Water Research*, 22(10), 1229–1238.

Hamoda, M.F., Al-Ghusain, I., Al-Mutairi, N.Z. (2004). Sand filtration of wastewater for tertiary treatment and water reuse. *Desalination*, 164(3), 203–211.

Hutchinson, W.R. (1976). High rate direct filtration. *Jour. Amer. Water Works Assoc.*, 65(9), 292–298.

Kawamura, S. (1975). Design and operation of high-rate filters – Part 1. *J. Amer. Water Works Assoc.*, 67(10), 535–549.

Koivunena, J., Siitonenb, A., Heinonen-Tanskia, H. (2003). Elimination of enteric bacteria in biological–chemical wastewater treatment and tertiary filtration units. *Water Research*, 37, 690–698.

Kozeny, J. (1927). Uber Kapillare Leitung des Wasser in Boden. *Wien Akad. Wiss.*, 136, 271–306.

Lawler, D.F. and Nason, J.A. (2006). Granular media filtration: Old process, new thoughts. *Water Science and Technology*, 53(7), 1–7.

Leva, M. (1959). *Fluidization*. McGraw Hill, New York.

Liao, C.-W., Yu, J.-H., Tang, Y.-S. (2010). On-line full scan inspection of particle size and shape using digital image processing. *Particuology, Chinese Society of Particuology*, 8(3), 286–292.

McEwen, B. (1998). Indirect potable reuse of reclaimed water. In *Wastewater Reclamation and Reuse*, Asano, T. (ed.). CRC Press, New York.

Mitchell, C., Chong, J., Turner, A., Retamal, M., Carrard, N., Murta, J., Mukheibir, P., Moy, C. (2013). Darling Quarter case study. Building industry capability to make recycled water investment decisions. Report, Institute for Sustainable Futures, University of Technology, Sydney, for the Australian Water Recycling Centre of Excellence.

Mujeriego, R., Cazurra, T., Gullon, M. (2008). Wastewater reuse in El Prat de Llobregat, Barcelona. *Water and Sciences Technology*, 57(4), 567–574.

Ncube, P., Pidou, M., Stephenson, T., Jefferson, B., Jarvis, P. (2016). The effect of high hydraulic loading rate on the removal efficiency of a quadruple media filter for tertiary wastewater treatment. *Water Research*, 107, 102–112.

Ni, C.H., Chen, J.N., Tsai, Y.C., Chen, T.K., Chen, W.B., Chen, C.H. (2003). Recycling and reuse of wastewater from a new-developed community using sand filtration, ultrafiltration, and ozonation. *J. Environ. Sci. Health Toxicol., Hazard. Subst. Environ. Eng.*, 38(10), 2339–2348.

Ogoshi, M., Suzuki, Y., Asano, T. (2001). Water reuse in Japan. *Water Sci. Technol.*, 3(10), 17–23.

Okun, D. (1997). Distributing reclaimed water through dual systems. *J. Amer. Water Works Assoc.*, 89(11), 52–64.

Olivieri, A.W., Eisenberg, D.M., Cooper, R.C., Tchobanoglous, G., Gagliardo, P. (1996). Recycled water – A source of potable water: City of San Diego health effects study. *Water Sci. Technol.*, 33(10/11), 285–296.

Perez, S., Köck-Schulmeyer, M., Ginebreda, A. (eds) (2010). Wastewater reuse in the mediterranean area of Catalonia, Spain: Case study of reuse tertiary effluent from a wastewater treatment plant at El Prat de Llobregat (Barcelona). In *Waste Water Treatment and Reuse in the Mediterranean Region*. Springer, Berlin.

Pollice, A., Lopez, A., Laera, G., Rubino, P., Lonigro, A. (2004). Tertiary filtered municipal wastewater as alternative water source in agriculture: A field investigation in Southern Italy. *Science of the Total Environment*, 24, 201–210.

Quinzaños, S., Strube, R., Dahl, C., Mujeriego, R. (2008). Helminth eggs removal by micro screening for water reclamation and reuse. *Water Science & Technology*, 57(5), 715–720.

Rodgers, S.E. and Lauer, W.C. (1992). Denver's demonstration of potable water reuse: Water quality and health effects testing. *Water Sci. Technol.*, 26(7/8), 1555–1564.

Sanz, J., Strube, R., Quinzaños, S., Dahl, C., Montoliu, I., Gracenea, M., Mujeriego, R. (2009). Helminth eggs removal in water reclamation on disk-filters. In *International Conference on Disinfection of Wastewater*. Universitat Politecnica Catalunya, Barcelona.

Sneed, E.D. and Folk, R.L. (1958). Pebbles in the Lower Colorado River, Texas: A study in particle morphogenesis. *Journal of Geology*, 66(2), 114–150.

Stefanakis, A. (2015). Modern water reuse technologies. In *Urban Water Reuse Handbook*, Eslamian, S. (ed.). CRC Press, New York.

Tchobanoglous, G., Burton, F.L., Stensel, H.D. (2002). *Wastewater Engineering: Treatment and Reuse*. Metcalf & Eddy, McGraw-Hill, New York.

Wen, C.Y. and Yu, Y.H. (1966). Mechanics of fluidization. *Chemical Engineering Progress*, 62(4), 75–85.

Yao, K.M., Habibian, M.T., O'Melia, C.R. (1971). Water and wastewater filtration: Concepts and application. *Environ. Sci. Technol.*, 5, 1105–1112.

Zhou, H. and Smith, D.W. (2002). Advanced technologies in water and wastewater treatment. *Journal of Environmental Engineering and Science*, 1, 247–264.

Disinfection

Antonelli, M., Rossi, S., Mezzanotte, V., Nurizzo, C. (2006). Secondary effluent disinfection: PAA long-term efficiency. *Environmental Science & Technology*, 40, 4771–4775.

Appels, L., Assche, A.V., Willems, K., Degrève, J., Impe, J.V., Dewil, R. (2011). Peracetic acid oxidation as an alternative pre-treatment for the anaerobic digestion of waste activated sludge. *Bioresource Technology*, 102, 4124–4130.

Azzellino, A., Antonelli, M., Canziani, R., Malpei, F., Marinetti, M., Nurizzo, C. (2011). Multivariate modelling of disinfection kinetics: A comparison among three different disinfectants. *Desalination and Water Treatment*, 29, 128–139.

Bach, R.D., Ayala, P.Y., Schlegel, H.B. (1996). A reassessment of the bond dissociation energies of peroxides. An ab initio study. *Journal of American Chemical Society*, 118, 12758–12765.

Baldry, MG.C., Cavadore, A., French, M.S., Massa, G., Rodrigues, L.M., Schirch, P.F.T., Threadgold, T.L. (1995). Effluent disinfection in warm climates with peracetic acid. *Water Science & Technology*, 31, 161–164.

Bianchini, R., Calucci, L., Caretti, C., Lubello, C., Pinzino, C., Piscicelli, M. (2002a). An EPR study on wastewater disinfection by peracetic acid, hydrogen peroxide and UV irradiation. *Annali di Chimica*, 92, 783–793.

Bianchini, R., Calucci, L., Caretti, C., Lubello, C., Pinzino, C. (2002b). Intermediate free radicals in the oxidation of wastewater. *Research on Chemical Intermediates*, 28, 247–256.

Bolton, J.R. and Karl, G.L. (2003). Standardization of methods for fluence (UV dose) determination in bench-scale UV experiments. *Journal of Environmental Engineering*, 29(3), 209–215.

Briancesco, R., Veschetti, E., Ottaviani, M., Bonadonna, L. (2005). Peracetic acid and sodium hypochlorite effectiveness in reducing resistant stages of microorganisms. *Central European Journal of Public Health*, 13, 159–162.

Bukhari, Z., Hargy, T.M., Bolton, J.R., Dussert, B., Clancy, J.L. (1999). Medium pressure UV light for oocyst inactivation. *J. Amer. Water Works Assoc.*, 91, 86–94.

Cavallini, G.S., De Campos, S.X., De Souza, J.B., Vidal, C.M. (2013b). Evaluation of the physico-chemical characteristics of wastewater after disinfection with peracetic acid. *Water, Air, & Soil Pollution*, 224, 1752–1763.

Chhetri, R.K., Thornberg, D., Berner, J., Gramstad, R., Öjstedt, U., Sharma, A.K., Andersen, H.R. (2014). Chemical disinfection of combined sewer overflow waters using performic acid or peracetic acids. *Science of the Total Environment*, 490, 1065–1072.

Chick, H. (1908). Investigations of laws of disinfection. *Journal of Hygiene*, 8(1), 92–158.

Clark, R.M. (1990). Modeling inactivation of *Giardia lamblia. Journal of Environmental Engineering*, 116(5), 837–853.

Collivignarelli, C., Bertanza, G., Pedrazzani, R. (2000). A comparison among different wastewater disinfection systems: Experimental results. *Environmental Technology*, 21(1), 1–16.

Craik, S.A., Weldon, D., Finch, G.R., Bolton, J.R., Belosevic, M. (2001). Inactivation of *Cryptosporidium parvum* oocysts using medium- and low-pressure ultraviolet radiation. *Water Res.*, 35, 1387–1398.

Darby, J.L., Snider, K.E., Tchobanoglou, G. (1993). Ultraviolet disinfection for wastewater reclamation and reuse subject to restrictive standards. *Wat. Env. Res.*, 65, 169–180.

Dell'Erba, A., Falsanisi, D., Liberti, L., Notarnicola, M., Santoro, D. (2007). Disinfection byproducts formation during wastewater disinfection with peracetic acid. *Desalination*, 215, 177–186.

Dorn, J.M. (1974). A comparative study of disinfection on viruses and bacteria by monochloramine. Master's Thesis, University of Cincinnati, Cincinnati.

Falsanisi, D., Gehr, R., Santoro, D., Dell'Erba, A., Notarnicola, M., Liberti, L. (2006). Kinetics of PAA demand and its implications on disinfection of wastewater. *Water Quality Research Journal of Canada*, 41, 398–409.

Féliers, C. and Laplace, C. (2001). Evaluation de la dose dans un réacteur vertical en utilisant la biodosimétrie, l'actinométrie et le calcul CFD. Report, 2nd World Water Congress, Berlin.

Finch, G.R., Haas, C.N., Oppenheimer, J.A., Gordon, G., Trussell, R.R. (2001). Design criteria for inactivation of cryptosporidium by ozone in drinking water. *Ozone: Science & Engineering. Journal of the International Ozone Association*, 23(4), 259–284.

Flores, M.J., Lescano, M.R., Brandi, R.J., Cassano, A.E., Labas, M.D. (2014). A novel approach to explain the inactivation mechanism of *Escherichia coli* employing a commercially available peracetic acid. *Water Science & Technology*, 69, 358–363.

Gaid, K. (2023). *Drinking Water Treatment 1–5*. ISTE Ltd, London, and John Wiley and Sons, New York.

Gaid, K., Girodet, P., Bord, G., Féliers, C. (2005). La désinfection aux ultraviolets : une alternative de choix. *L'eau, l'industrie, les nuisances*, 282, 45–52.

Gehr, R. and Cochrane, D. (2002). Peracetic acid (PAA) as a disinfectant for municipal wastewater: Encouraging performance results from physicochemical as well as biological effluents. In *Proceedings of the US Water Environment Federation Disinfection Conference*. Water Environment Federation, 182–198.

Gehr, R., Wagner, M., Veerasubramanian, P., Payment, P. (2003). Disinfection efficiency of peracetic acid, UV and ozone after enhanced primary treatment of municipal wastewater. *Water Research*, 37, 4573–4586.

Gehr, R., Chen, D., Moreau, M. (2009). Performic acid (PFA): Tests on an advanced primary effluent show promising disinfection performance. *Water Science & Technology*, 59, 89–96.

Gyürék, L.L. and Finch, G.R. (1998). Modeling water treatment chemical disinfection kinetics. *Journal of Environmental Engineering*, 124, 783–793.

Haas, C.N. and Engelbrecht. R.S. (1980). Chlorine dynamics during inactivation of coliforms, acid-fast bacteria and yeasts. *Water Res.*, 14, 1744–1755.

Heinonen-Tanski, H. and Miettinen, H. (2010). Performic acid as a potential disinfectant at low temperature. *Journal of Food Process Engineering*, 33(6), 1159–1172.

Hill, A.G. and Rice, R.G. (1982). Historical background, properties and applications. In *Handbook of Ozone Technology and Applications*, Rice, R.G. and Netzer, A. (eds). Ann Arbor Science Publishers, Ann Arbor.

Jacangelo, J., Darby, J.L., Loge, F., Tchobanoglous, G., Heath, M., Swaim, P. (1995). Comparison of UV irradiation to chlorination: Guidance for achieving optimal UV performance. Final Report, Water Environment Research Foundation, Project 91-WWD-1.

Karpova, T., Pekonen, P., Gramstad, R., Öjstedt, U., Laborda, S., Heinonen-Tanski, H., Chávez, A., Jiménez, B. (2013). Performic acid for advanced wastewater disinfection. *Water Science & Technology*, 68, 2090–2096.

Kitis, M. (2004). Disinfection of wastewater with peracetic acid: A review. *Environ. International*, 30, 47–55.

Koivunen, J. and Heinonen-Tanski, H. (2005a). Inactivation of enteric microorganisms with chemical disinfectants, UV irradiation and combined chemical/UV treatment. *Water Research*, 39, 1519–1526.

Koivunen, J. and Heinonen-Tanski, H. (2005b). Peracetic acid (PAA) disinfection of primary, secondary and tertiary treated municipal wastewater. *Water Research*, 39, 4445–4453.

Koivunen, J., Siitonen, A., Heinonen-Tanski, H. (2003). Elimination of enteric bacteria in biological–chemical wastewater treatment and tertiary filtration units. *Water Research*, 37, 690–698.

Kollu, K. and Ormeci, B. (2012). Effect of particles and bioflocculation on ultraviolet disinfection of Escherichia coli. *Wat. Res.*, 46, 750–760.

Lazarova, V., Janex, M.L., Fiksdal, L., Oberg, C., Barcina, I., Pommepuy, M. (1998). Advanced wastewater disinfection technologies: Short and long term efficiency. *Water Science & Technology*, 38(12), 109–117.

Lefevre, F., Audic, J.M., Ferrand, F. (1992). Peracetic acid disinfection of secondary effluents discharged off coastal seawater. *Water Science & Technology*, 25, 155–164.

Li, H., Finch, G.R., Smith, D.W., Belosevic, M. (2001). Chlorine dioxide inactivation of *Cryptosporidium parvum* in oxidant demand-free phosphate buffer. *Journal of Environmental Engineering*, 127(7), 594–603.

Liberti, L. and Notarnicola, M. (1999). Advanced treatment and disinfection for municipal wastewater reuse in agriculture. *Water Science & Technology*, 40(4/5), 235–246.

Liberti, L., Lopez, A., Notarnicola, M., Barnea, N., Pedahzur, R., Fattal, B. (2000). Comparison of advanced disinfecting methods for municipal wastewater reuse in agriculture. *Water Science & Technology*, 42, 215–220.

Luukkonen, T. (2016). New adsorption and oxidation bases approaches for water and wastewater treatment. PhD Thesis, University of Oulu, Kuusamonsali.

Luukkonen, T., Teeriniemi, J., Prokkola, H., Rämö, J., Lassi, U. (2014). Chemical aspects of peracetic acid based wastewater disinfection. *Water SA*, 40, 73–80.

Luukkonen, T., Heyninck, T., Lassi, U., Rämö, J. (2015). Comparison of organic peracids in wastewater treatment: Disinfection, oxidation and corrosion. *Water Research*, 85, 275–285.

Mazzini, R., Pedrazzi, L., Lazarova, V. (2013). Production d'eau recyclée de haute qualité pour l'irrigation agricole à Milan. In *Jalons de la réutilisation de l'eau : les meilleures histoires de réussite*, Lazarova, V., Asano, T., Bahri, A., Anderson, J. (eds). IWA Publishing, London.

McNamara, P.J., Tucker, R., Burch, T.R., LaPara, T.M., Tan, D.T., Yan, M. (2011). Tertiary-treated municipal wastewater is a significant point source of antibiotic resistance genes. *Environmental Science & Technology*, 45(22), 48–62.

Merka, V., Sita, F., Zikes, V. (1965). Performic and perpropionic acids as disinfectants in comparison with peracetic acid. *Journal of Hygiene, Epidemiology, Microbiology, and Immunology*, 59, 220–226.

Mezzanotte, V., Antonelli, M., Citterio, S., Nurizzo, C. (2007). Wastewater disinfection alternatives: Chlorine, ozone, peracetic acid, and UV light. *Water Environment Research*, 79(12), 2373–2379.

Ragazzo, P., Chiucchini, N., Bottin, F. (2007). The use of hyproform disinfection system in wastewater treatment: Batch and full scale trials. In *Chemical Water and Wastewater Treatment IX*, Hahn, H.H., Hoffmann, E., Odegaard, H. (eds). IWA Publishing, London.

Ragazzo, P., Chiucchini, N., Piccolo, V., Ostoich, M. (2013). A new disinfection system for wastewater treatment: Performic acid full scale trials evaluations. *Water Sci. Technol.*, 67(11), 2476–2487.

Rajala-Mustonen, R.L., Toivola, P.S., Heinonen-Tanski, H. (1997). Effects of peracetic acid and UV irradiation on the inactivation of coliphages in wastewater. *Water Science & Technology*, 35, 237–241.

Rossi, S., Antonelli, M., Mezzanotte, V., Nurizzo, C. (2007). Peracetic acid disinfection: A feasible alternative to wastewater chlorination. *Water Environment Research*, 79, 341–350.

Santoro, D., Gehr, R., Bartrand, T.A., Liberti, L., Notarnicola, M., Dell'Erba, A., Falsanisi, D., Haas, C.N. (2007). Wastewater disinfection by peracetic acid: Assessment of models for tracking residual measurements and inactivation. *Water Environment Research*, 79, 775–787.

Shin, G.A., Linden, K.G., Sobsey, M.D. (2005). Low pressure ultraviolet inactivation of pathogenic enteric viruses and bacteriophages. *J. Environ. Engr. Sci.*, 4, 7–11.

Swern, D. (1949). Organic peracids. *Chemical Reviews*, 45, 1–68.

Veschetti, E., Cutilli, D., Bonadonna, L., Briancesco, R, Martini, C., Cecchini, G., Anastasi, P., Ottaviani, M. (2003). Pilot-plant comparative study of peracetic acid and sodium hypochlorite wastewater disinfection. *Water Research*, 37, 78–94.

Watson, H.E. (1908). A note on the variation of the rate of disinfection with change in the concentration of the disinfectant. *Journal of Hygiene*, 8(4), 536–542.

Yousefzadeh, S., Nabizadeh, R., Mesdaghinia, A., Nasseri, S., Hezarkhani, P., Beikzadeh, M., Valadi, A.M. (2014). Evaluation of disinfection efficiency of performic acid (PFA) catalyzed by sulfuric and ascorbic acids tested on *Escherichia coli* (ATCC, 8739). *Desalination and Water Treatment*, 52, 3280–3289.

Zuccato, E., Castiglioni, S., Mazzini, R. (2009). Pollution de l’environnement par les résidus de médicaments. *Ingengneria ambiental*, 7(8), 353–357.

2

Micropollutants

2.1. Introduction

Numerous tertiary treatment options have been examined for the reuse of wastewater for non-drinking purposes. Some applications require technical mastery and vigilance, for example, when used for irrigating raw foods.

For drinking water applications, more effective treatments are needed to remove micropollutants, salts and, more importantly, pathogens. Not all these requirements can be fully met by the tertiary treatments described above.

To this end, a new stage consisting of more efficient processes must be included in the treatment process for drinking water applications.

The aim of quaternary treatment is to minimize the presence of emerging pollutants (antibiotics, pesticides, hormones, anticancer agents, etc.) and microbiology in reused water, the presence of which in wastewater is certain to hinder its use as drinking water.

The removal of pesticides, endocrine disruptors (EDs), drug residues and pathogens is therefore necessary, involving advanced chemical oxidation processes or the integration of membranes into the treatment process.

These processes are as follows:

– ozonation, alone or combined with activated carbon;

– advanced oxidation (AOP) combining various chemical and physical oxidants;

– activated carbon adsorption (powdered activated carbon (PAC) or granular activated carbon [GAC]);

– membranes (microfiltration, ultrafiltration, nanofiltration and reverse osmosis).

These treatments are well known, as most of them have already been tried and tested, and are used to treat drinking water.

Given that aquatic environments have always been intimately linked to economic development, including population growth, industrial expansion, intensive agriculture and mass tourism, they have become the ultimate dumping ground for the waste associated with our lifestyles. Micropollutants in the aquatic environment are now a major problem both for the human population using water resources and for aquatic ecosystems.

Micropollutants in wastewater are made up of numerous mineral and organic compounds that can be toxic to humans and/or aquatic organisms, even at very low concentrations in the water (from nanograms to a hundred micrograms per liter).

These micropollutants come from a variety of sources (Table 2.1):

– economic activities (plasticizers, flame retardants, electrical insulators, etc.);

– agricultural use (e.g. use of plant protection products and veterinary medicines);

– domestic use (cleaning products, detergents, cosmetics, medicines, etc.);

– induced contamination from atmospheric deposition and water runoff on urban surfaces;

– natural origin (metals derived from the geology of the environment).

2.2. Pesticides

Agriculture is by far the largest consumer of pesticides. The majority of active substances emitted into the environment are used by agriculture to control pests and weeds.

The list of biocides reveals three main families: insecticides, fungicides and herbicides. Insecticides can affect the nervous system (neurotransmission), respiration or molting.

Main families	Compounds
Medicinal molecules	Anxiolytics, analgesics, anti-inflammatories, antibiotics, hypolipidemics, antihypertensives, antiepileptics, anticancer agents, iodinated contrast agents, etc.
Cosmetics and personal hygiene products	Musks (galaxolide and tonalide), triclosan, etc.
Endocrine disruptors	Hormones, alkylphenols, BPA, phthalates, flame retardants, etc.
Illicit drugs	Cocaine and metabolites, THC, amphetamines, etc.
Pesticides	Insecticides and herbicides
Priority substances (Appendix X of the WFD) and hazardous substances (circular RSDE 050109)	PAHs and dichloromethane
Disinfection by-products	Trihalomethanes and haloacetic acids

Table 2.1. *Main families and organic compounds*

Fungicides act selectively on respiration, biosynthesis (sterols, RNA and melanin), growth or cell permeability. Lastly, herbicides can have a detrimental impact on photosynthesis, lipid and amino acid synthesis, cell division or growth. The most common pesticide families include:

– organophosphates (chlorpyrifos and malathion), which have an insecticidal effect by blocking acetylcholinesterase (disrupting the transmission of nerve impulses);

– organochlorines (DDT, chlordecone and lindane), insecticides that act on the nervous system by disrupting nerve impulses at synapses. These compounds are now all banned in France, but are very persistent in the environment;

– carbamates (aldicarb and carbofuran), which act on the nervous system in the same way as organophosphates;

– triazines (atrazine and simazine), which have a herbicidal effect by blocking electron transport and light energy;

– phenylureas (diuron and isoproturon), which have a mode of action comparable to that of triazines;

– chloroacetanilides (metolachlor and acetochlor), which act on plants by inhibiting elongases, enzymes involved in lipid synthesis (long-chain fatty acids);

– aryl-oxyacids (2,4 D, mecoprop and dichlorprop), which interfere with the regulation of auxin AIA (indole 3-acetic acid), a phytohormone that is essential for plant growth;

– sulfonylureas (chlorsulfuron and nicosulfuron), which act on the synthesis of branched amino acids (valine, leucine and isoleucine) by inhibiting acetolactate synthase.

Farming is not the only source of pesticide emissions. Secondary sources include the upkeep of public spaces, golf courses, use by private individuals, annual mosquito control in wetlands, the veterinary sector, road maintenance, etc.

2.3. Pharmaceuticals and industrial residues

France is one of the world's biggest consumers of medicines. Over 3,000 pharmaceutical molecules for human use are currently used in medicines, and 300 veterinary drugs are currently available on the French market. The biologically active substances contained in each pharmaceutical specialty are characterized by a wide diversity of chemical structures. Drug residues mainly are excreted in feces and urine in their original form or in the form of one or more active or inactive metabolites, and then released into sewage systems, wastewater and, in the case of most veterinary drugs, into the soil.

The main sources of water pollution by pharmaceuticals are diffuse sources, mainly from human and animal excretions, which account for most of the drugs discharged into the environment, and point sources, which are responsible for much more concentrated but geographically limited emissions.

Diffuse sources include inappropriate storage and disposal of expired or unused drugs, and metabolic excretion through the urinary or digestive tracts of drugs consumed by humans and animals.

Drug substances are grouped into several classes, the latter comprising a wide range of chemical families, as well as degradation metabolites produced by people or in the environment:

– analgesics and anti-inflammatories, which are the most widely consumed molecules in France for human use, such as diclofenac and ibuprofen, acetylsalicylic acid and paracetamol;

– psychotropic drugs, including antidepressants, anxiolytics and sleeping pills;

– anticonvulsants;

– lipid-lowering agents of the fibrate family, designed to lower plasma cholesterol and triglyceride concentrations;

– beta-blockers, which reduce blood pressure by reducing the binding of adrenalin to its receptors. Both are cardioselective, since they act specifically by reducing nerve impulses to the heart;

– antibiotics, designed to destroy or block the action of bacteria, for example, by disrupting the synthesis of proteins associated with bacterial growth;

– bronchodilators, which act by stimulating beta-2 receptors in the bronchial tubes, causing them to dilate in the short term.

Other classes include diuretics, antifungals, antivirals and laxatives.

Furthermore, the use of different antibiotics for animal production and human medicine may explain the observation of distinct correlations between bacterial community composition and antibiotic-resistant populations, and the concentration of tetracyclines, sulfonamides and penicillins or fluoroquinolones in wastewater. In fact, antibiotic-resistant bacteria have been detected in several sources of wastewater and treated wastewater.

Point sources include the direct discharge of drugs (and chemical products used in drug manufacture) into wastewater from the pharmaceutical and chemical industries, and the direct or indirect removal of pharmaceutical agents used in healthcare establishments, the direct dispersion of veterinary drugs into the environment in aquaculture or during the treatment of farm animals, or indirect dispersion when liquid manure is spread on soils intended for agricultural use, and the direct dispersion of therapeutic molecules in the form of food supplements poured directly into fish-farming ponds.

These substances are found in various environmental compartments (water, air, soil), through direct or indirect emissions (runoff, drainage, atmospheric fallout, etc.), with potential direct or indirect effects on human health and ecosystems, notably through contamination of the trophic chain (Figure 2.1). These molecules are characterized by a wide variety of chemical structures. When molecules used for therapeutic or diagnostic purposes are not completely degraded in the body, they are mainly excreted through feces and urine in their initial form or in the form of one or more metabolites. These human or veterinary drug residues may, after several stages, be re-released into the natural environment. Concentration levels in natural waters vary according to chemical stability, biodegradability, the physicochemical characteristics of the molecules and the performance of wastewater treatment plants.

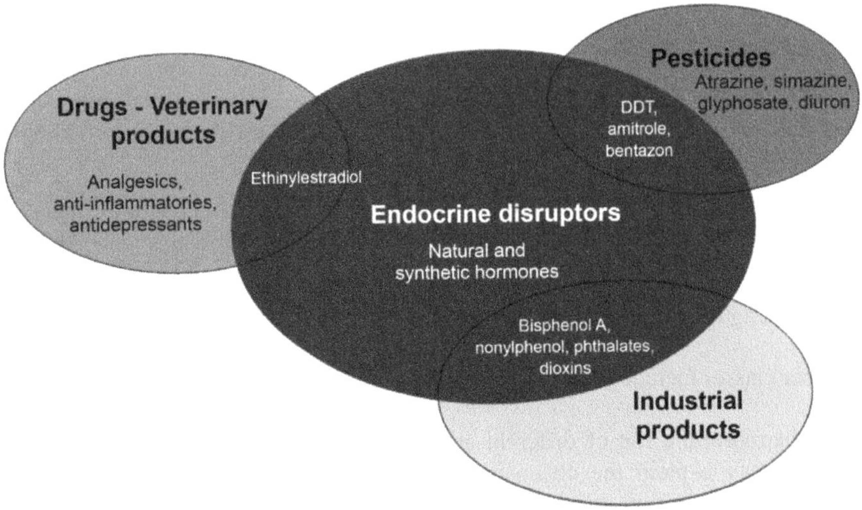

Figure 2.1. *Different categories of micropollutants. For a color version of this figure, see www.iste.co.uk/gaid/reuse2.zip*

Generally speaking, the concentrations of pesticides and pharmaceuticals found in the environment not only vary from one compound to another, but also from one site to another. Regional differences in drug use and the biodegradability of pharmaceuticals and their metabolites could explain these differences in concentrations.

Analysis campaigns carried out at the outlets of wastewater treatment plants show the presence of these substances, often in significant concentrations.

Due to the diversity and complexity of the substances considered, data are still limited on the effects of pharmaceutical residues and EDs in the environment, as well as on their degree of occurrence in resources. Given the very wide range of molecules considered and the difficulties involved in their analysis, which have sometimes been linked to their low concentrations, the studies carried out to date to assess the presence of these compounds in the environment have been limited to a few groups of molecules. These have been selected on the basis of production quantities and estimates of their persistence in the environment in the case of pharmaceutical substances residues (PSR), and of their potential endocrine activity in the case of EDs. Some of these molecules act on the hormonal balance of numerous living species and are therefore qualified as EDs. Recent toxicological studies have shown that, in humans, EDs can affect sperm count, birth defects, sexual function and neuronal development, obesity and even cancer.

	0.01–0.1	0.1–1.0	1.0–10	10–100
> 70 %	Terbutaline, Diazepam, Salbutamol, Bromazepam, 17 Estradiol, Estrone, Fluoxetine, Nordazepam	Amitriptyline, Metoprolol, Roxithromycin, Bisoprolol, Propranolol, Estriol, Carbamazepine, Diclofenac, Gemfibrozil, Sotalol, Sulfamethoxazole Diuron, Fluoranthene	Naproxen, Atenolol, Acetobutolol, Caffeine, Ketoprofen	46 DCE substances
30 %–70 %	Atrazine, Simazine, DEA, DIA, Tributyltin, Pentachlorophenol Indeno(1,2,3)pyrene, Endosulfan, Isoproturon, Benzol (g, h, i)perylene, Isodrine, CCIA	Dichlorophenol, Benzothiazole, 4-tert-butylphenol Naphthalene, Trichlorobenzene, Benzo(b)fluoranthene AMPA, Triclosan, Tributylphosphate	Di- and trichloromethane, Tetrachloroethylene Decabromodiphenyl ether	
30 %	1-dichloroethane, Anthracene, Benzo(a)pyrene, Benzene, Dieldrin, Hexachlorobenzene, Hexabromodiphenyl ether, Di- and tribromophenol, Tri- and tetrachlorophenol	Benzo(k)fluoranthene, Chlorpyrifos, Hexachlorocyclohexane, Trichlorethylene, Pentabromodiphenyl ether, Monochlorophenol, Glyphosate, Tetrabromodiphenylene	C10-C13 chloroalkanes, Tribromodiphenyl ether	

Concentration (µg.L^{-1})

Table 2.2. *Example of an assessment of micropollutant concentrations in water*

Substance	Concentration µg.L^{-1}	Substance	Concentration µg.L^{-1}
Acetobutolol	0.3–5.9	Galaxolide	0.1–1.4
Atenolol	0.15–3.2	Gemfibrozil	0.3–1.2
Clofibric acid	0.9	Ibuprofen	0.4–3.5
Bromazepam	0.02–0.12	Iopromide	2–2.9
Carbamazepine	0.1–0.9	Ketoprofen	0.2–3.1
Ciproflaxine	0.05–1.4	Metoprolol	0.2–1.6
Cycloproconazole	0.2–1.9	Naproxen	0.5–4
Diclofenac	0.9–1.2	Nordazepam	0.05–1.1
Estrone	0.05–1.1	Propranolol	0.6–3.8
17-β-Estradiol	0.2–1.4	Roxithromycin	0.05–1.5
17-α-Estradiol	0.2–1.2	Sulfamethoxazole	0.5–1.3
17-α-Ethinylestradiol	0.2–1.2	Sotalol	0.3–1.5
Estriol	0.05–1.1	Terbutaline	0.05–0.1
Erythromycin	0.2–2.9	Triadimefon	0.3–4.2
Fluoxetine	0.3–2.1	Tebuconazole	0.1–3.8

Table 2.3. *Examples of micropollutant concentrations in treated wastewater*

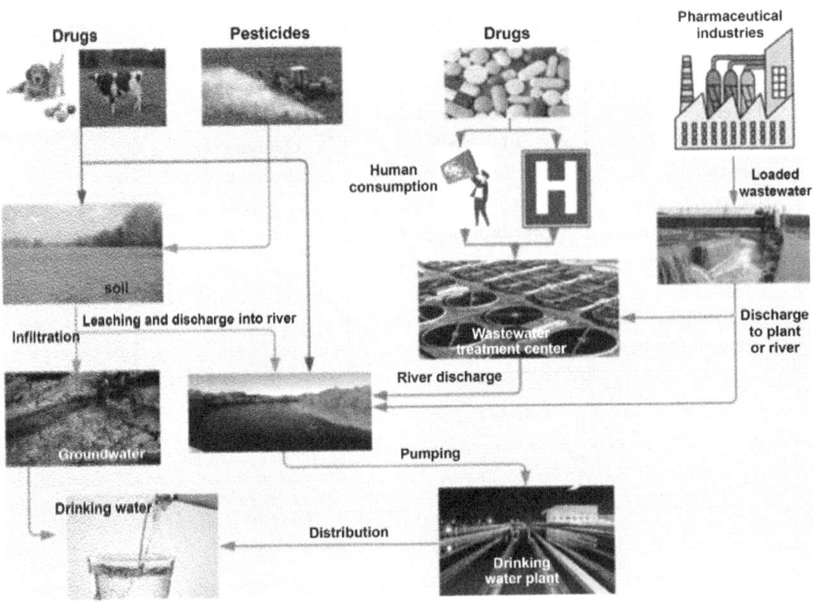

Figure 2.2. *Simplified diagram of the water cycle containing pesticides, drug residues and endocrine disruptors. For a color version of this figure, see www.iste.co.uk/gaid/reuse2.zip*

There is now little doubt about the harmful effects of environmental releases of EDs and PSRs on the populations of aquatic and terrestrial organisms (from simple gastropods to mammals). Numerous examples demonstrate the impact of these substances on fish (e.g. incomplete development of genital organs, hermaphroditism, altered male/female ratios), birds (e.g. thinning of egg shells) and mammals (e.g. decline in seal populations feeding on polychlorobiphényles (PCB)-contaminated fish). There are still many questions in the scientific world concerning the cause-and-effect relationship, which is often difficult to establish (still incomplete knowledge of the toxic and ecotoxic effects of metabolites, substance cocktails, disruption mechanisms, etc.).

According to the European Commission, an ED is “an exogenous substance or mixture which alters the functions of the endocrine system and thereby induces adverse effects on the health of an intact organism, its descendants or sub-populations”.

In humans, the links between the presence of pharmaceutical residues and EDs in the environment are a matter of public concern. European authorities seem to have accepted that an association between EDs and increasingly frequent pathologies, such as testicular, breast and prostate cancer, reduced sperm count, reproductive organ malformations and thyroid disorders, cannot be confirmed.

Their massive use, as well as the waste generated during their manufacture or use, results in their presence at concentration levels ranging from a few dozen nanograms per liter to a few micrograms per liter.

The presence of drug residues in drinking water is a major issue for the scientific community and public authorities. For several years now, the scientific community, public authorities and the general public have been concerned with the presence of trace amounts of drug residues in aquatic environments (surface water, groundwater, transfer media, soil) and in drinking water, and about their effects on the environment and human health. This concern is part of a more general context of preserving the environment and water resources, in particular against micropollutants, substances that are likely to have a toxic effect at low doses in a given medium (ranging from nanograms to micrograms per liter of water).

The French government has made source reduction a priority, even to the point of eliminating emissions of certain substances altogether. Source reduction measures are essential to prevent the release of substances into the environment, but they will take many years to become effective, and will not be sufficient to control all the risks associated with micropollutants. This is due to the very large number of molecules, uses and practices involved, but also due to the fact that many micropollutants are primarily the result of diffuse pollution emitted by individuals (drug residues, hormones, biocides, etc.), which will therefore be difficult to eliminate entirely.

In Europe, a regulatory framework has been put in place to reduce the discharge of hazardous substances into the natural environment and to study their outcome. The many efforts being made, particularly at the European level, are aimed at gaining a better understanding of the environmental (chemical) quality of the various water resources available, as well as their development.

In France, regulatory requirements aimed at reducing the emissions of chemical substances and their presence in the natural environment derive from European regulations and directives such as the REACH regulation, as well as the regulations on biocidal and phytopharmaceutical products, the Water Framework Directive (WFD), the Industrial Emissions Directive (IED) and so on. The main aim of these regulations is to control the marketing of potentially hazardous substances, reduce their emissions into the various environmental compartments and meet environmental quality objectives. These increasingly strict requirements will tend to become more stringent in the years to come, as scientific knowledge evolves and impacts on health and the environment come to light.

In addition to transposing these fundamental European texts into French law, the government has set up a micropollutant plan to preserve water quality and biodiversity. The plan aims to reduce emissions of micropollutants of proven toxicity, consolidate knowledge and draw up lists of pollutants on which to take action.

Municipal and industrial wastewater treatment has come a long way in recent decades, discharging effluent according to increasingly stringent water quality requirements. In the context of water reuse, unplanned indirect drinking water is a common practice worldwide.

However, with regard to micropollutants, which are emerging contaminants, monitoring studies of their presence in surface water or wastewater are being increasingly reported in the literature.

Although wastewater treatment plants have improved significantly in technical terms, they were not designed to remove all pharmaceutical molecules. For example, although paracetamol is more than 90% degraded, traces of it are still found in treated wastewater discharged into surface waters. Some compounds, such as diclofenac (an anti-inflammatory) or carbamazepine (an anti-epileptic), degrade very little.

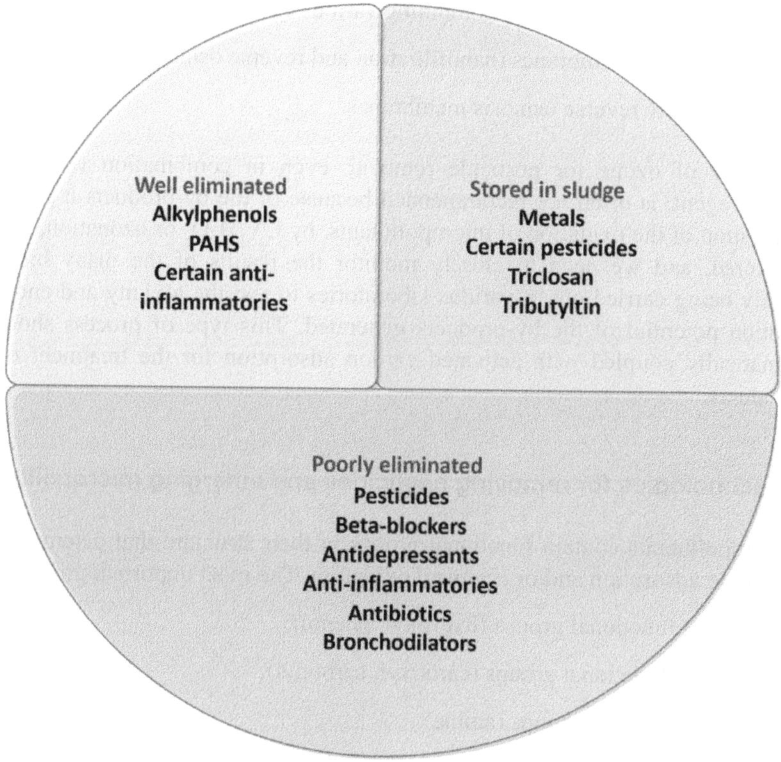

Figure 2.3. *Outcome of some substances in wastewater treatment plants*

Pharmaceutical residues discharged into surface waters can therefore present a significant environmental risk, as can their metabolites or degradation by-products, which are sometimes more dangerous than the original product.

Today's challenge is to remove micropollutants and microplastics from wastewater, especially if the wastewater is to be reused for drinking water purposes (urban and industrial). Consequently, additional treatment processes are required. Some plants already operate with these additional treatments (known as quaternary treatments).

Micropollutants can be removed using a variety of processes:

– activated carbon adsorption (PAC or GAC);

– ozonation, alone or combined with activated carbon;

– advanced oxidation (AOP) combining various chemical and physical oxidants;

– high-pressure membranes (nanofiltration and reverse osmosis);

– low-pressure reverse osmosis membranes.

The use of ozone for pesticide removal, even in combination with another oxidizing agent, is often not recommended because of the by-products it generates. The question of the oxidation of micropollutants, by UV/H_2O_2 or ozonation, remains unanswered, and we need to closely monitor the results of the many bioassays currently being carried out in various laboratories to test the toxicity and endocrine disruption potential of the by-products generated. This type of process should be systematically coupled with activated carbon adsorption for the treatment of any by-products.

2.4. Technologies for removing pesticides and emerging micropollutants

Micropollutants contain functional groups in their structure that determine their reactivity to adsorption and/or chemical oxidation. The most important groups are:

– OH-type functional groups (hydroxyl, phenol);

– C-O-type functional groups (carboxyl, carbonyl);

– N-type functional groups (amine).

Reactivity is strongly influenced by the polarity of functional groups and the acquisition of a positive or negative charge when proton dissociation or association takes place in these functional groups.

The *hydroxyl group* occurs in alcohols and phenols. Alcohols are considered to be hydroxylated alkenes. In most cases, alcohols are neutral, since the hydroxyl group does not ionize easily. Phenols contain hydroxyl components that are attached to the aromatic ring.

The *phenol group* consists of a hydroxyl group attached directly to the carbon of an aromatic ring. In general, phenols are highly reactive.

The *carbonyl group* includes aldehydes, ketones and carboxylic acids. Due to the double bond between carbon and oxygen, as well as the fact that the electrons in this double bond are not evenly distributed, the carbonyl group is polarized. Carbonyls acquire protons. In general, aldehydes are more easily oxidized than ketones.

The *carboxyl group* forms the weak acid group and incorporates the reactive properties of the carbonyl and hydroxyl groups.

The *amine group* is found in amines, which are usually organic bases that can be aliphatic or aromatic.

As a result, the presence of these groups influences the effectiveness of the treatment and can be a factor in the selection of the process to be implemented.

The treatment techniques presented in this chapter concern:

– simple or advanced chemical oxidation;

– activated carbon adsorption;

– combinations of chemical oxidation and activated carbon adsorption;

– membrane barriers (nanofiltration or low-pressure reverse osmosis);

– adapted biological treatments.

2.4.1. *Conventional oxidizers: chlorine, hypochlorite and chlorine dioxide*

The chemical oxidants commonly used in wastewater, such as chlorine, chlorine dioxide and hypochlorite, can be attractive reagents for the oxidation of micropollutants.

Chlorine and hypochlorite give interesting results for the elimination of certain organic micropollutants, with the exception of compounds with a ketone function conjugated to a double bond, which inhibit oxidation mechanisms. The highest performances were recorded for the tetracycline and diuretic antibiotics, with over 90% elimination. Conversely, psychiatric drugs and antihypertensives are less sensitive to chlorine, with less than 30% elimination. Opiates and cannabinoids showed good elimination (>90%) during chlorination. In contrast, cocaine and amphetamine-type compounds showed less than 40% elimination. Alkylphenols, nonylphenol and nonylphenol ethoxylates, which are xenobiotics, showed elimination rates in excess of 50%.

Despite interesting results, chlorine and hypochlorite should be considered carefully, as they produce a large number of by-products.

Chlorine dioxide, ClO_2, is a stable oxidant whose main reduction product is chlorite. Sulfamethoxazole, roxithromycin, 17-α-ethinylestradiol and diclofenac showed good degradation with ClO_2. This oxidant gives more interesting results than chlorine, but weaker than ozone due to its slower kinetics.

2.4.1.1. *Ozonation*

Ozone, a molecule made up of three oxygen atoms (in a gaseous state under normal temperature and pressure conditions), is a powerful oxidizing agent that transforms and degrades numerous organic substances. It is used on a large scale to remove tastes, odors and color, as well as to disinfect drinking water. Ozone is electrophilic and reacts rapidly and mainly selectively with amines, phenols, aromatic compounds and double bonds.

Oxidizer	Reactions	E° (V)
OH·	$2O_3 + H_2O_2 \rightarrow 2OH\cdot + 3O_2$	2.8
Ozone	$O_3 + 2H^+ + 2e^- \rightarrow O_2 + H_2O$ $O_3 + H_2O + 2e^- \rightarrow O_2 + 2OH^-$	2.08 1.24
Hypochlorite	$HClO^- + H^+ + 2e^- \rightarrow Cl^- + H_2O$ $ClO^- + H_2O + 2e^- \rightarrow Cl^- + 2OH^-$	1.48 0.84
Chlorine	$Cl_2 + 2e^- \rightarrow 2Cl^-$	1.36
Chlorine dioxide	ClO_2 (aqueous) + $e^- \rightarrow ClO_2^-$	0.95

Table 2.4. *Redox potentials of some conventional chemical oxidizers*

Due to its high oxidation potential, ozone treatment is widely used in a wide range of micropollutant removal applications, including pharmaceuticals, EDs and pesticides.

Thus, ozone reacts with organic contaminants both through a direct reaction with molecular ozone and through indirect reactions with free radicals (including the hydroxyl radical OH·), produced by ozone decomposition. Indeed, the decomposition of ozone in water and its reaction with organic matter lead to the formation of the hydroxyl radical (OH·), another powerful oxidant that reacts very rapidly and non-selectively with most organic molecules. The rate of OH· formation depends on the water matrix, in particular its pH, alkalinity, type and content of natural organic matter. It can significantly improve the oxidation of certain micropollutants. The reactivity of a substance with ozone depends on its molecular structure and can be characterized by its second-order reaction rate constants k_{O3} (with ozone) and $k_{OH\cdot}$ (with the hydroxide radical). The reaction kinetics accompanying chemical ozone oxidation mechanisms take place in ozonation tanks (two to four chambers), into which O_3 is injected (Figure 2.4). Contact times in each chamber can range from 4 to 8 min. Reactor design and the composition of the water to be treated are important parameters influencing O_3 oxidation performance.

Ozone oxidation reactions are generally favored at high pH levels due to the increased production of hydroxyl radicals. Pharmaceutical products (and xenobiotics in general) can be transformed into composites that are even more toxic than the parent molecules, even with low doses of ozone.

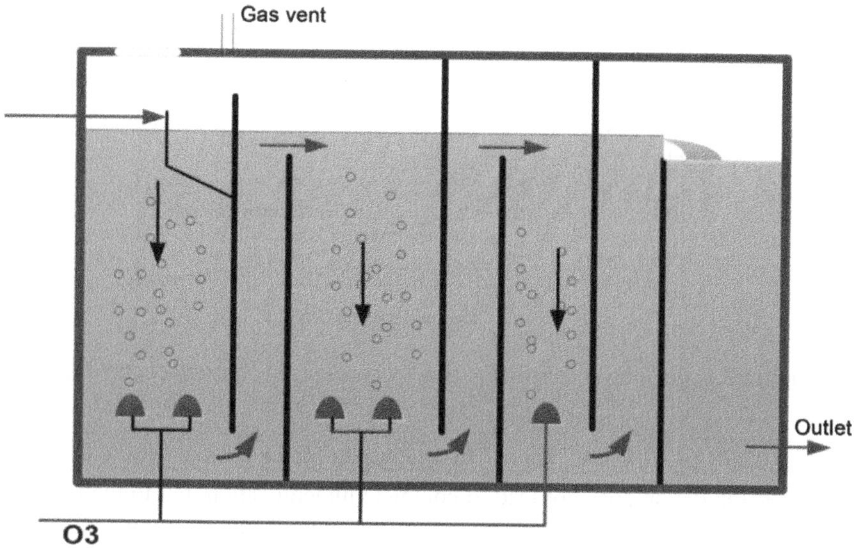

Figure 2.4. *Three-tank ozonation reactor*

The higher the ozone concentration and the longer the exposure time, the better the elimination of pollutants. Ozone not only reacts with micropollutants, but also with organic matter in general, including nitrites and other inorganic compounds. In addition to increasing ozone consumption, this also entails the risk of undesirable reaction by-products such as nitrosamines, bromates, etc., being formed, as many oxidized substances are generally not completely mineralized, and by-products are formed.

2.4.1.1.1. Pesticides

Ozone reacts rapidly with double bonds, deprotonated amines and activated aromatic compounds, which is often the case with pesticides and other micropollutants such as drug residues and synthetic substances. At $pH > 8$ in particular, ozonation is considered an advanced oxidation process due to the decomposition of O_3 molecules into a high concentration of $OH\cdot$ radicals, which involves interactions between these radicals and organic molecules.

Pesticide reaction rate constants with ozone range from <0.04 $M^{-1}.s^{-1}$ for lindane to 4.4 × 10^4 for aldicarb. Lindane reacts very slowly, while aldicarb reacts in a matter of seconds.

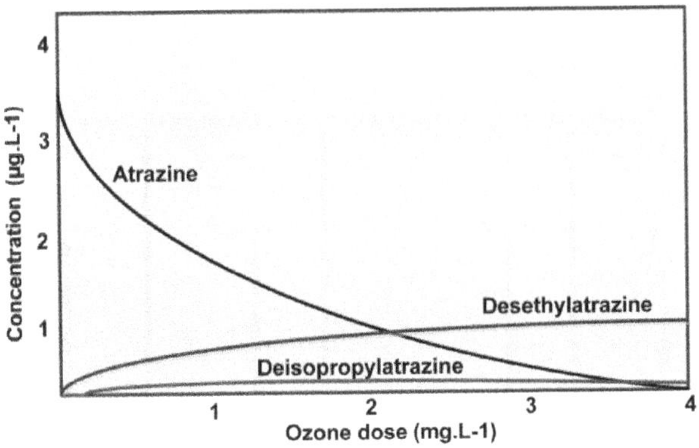

Figure 2.5. *Ozonation of atrazine and by-products at different ozone doses*

Reactions with pesticides are varied, depending on the chemical properties of the molecules. Oxidation does not lead to complete mineralization, but to the formation of oxidation by-products through protonation, dimerization, isomerization and tautomerization reactions. The rate constants of atrazine, cyanazine and deisopropylatrazine (DIA) are very similar. Atrazine has a removal rate of less than 50% with an ozone dose of 2.5 $mg.L^{-1}$, but this is significantly improved using advanced oxidation processes (UV/O_3, O_3/H_2O_2) (see section 2.4.2). DEA and DIA, which are by-products of the chemical oxidation of atrazine, and the herbicide cyanazine, are rather recalcitrant molecules to oxidize, as they do not contain an aromatic function, but instead an electron-withdrawing group (chlorine double bond), which leads to this low reactivity towards ozone. Generally speaking, pesticides are more difficult to oxidize than drug residues with, for example, less than 65% removal for cyanazine with a Ct of 13.2 $mg.min.L^{-1}$. Removals for DEA and DIA are around 38–50% for the same Ct.

River water containing approximately 2–2.5 mgL^{-1} dissolved organic carbon (DOC) and 0.3–0.4 μgL^{-1} atrazine is treated by coagulation-flocculation sand filtration. A complementary treatment by ozone shows a complete degradation of this pesticide for a dose of 4 $mg.L^{-1}$ and a contact time of 6 min. For a dose of 2 $mg.L^{-1}$, the atrazine residual is around 0.8 $\mu g.L^{-1}$. The DEA and DIA by-products are rapidly present at doses below 0.5 $mg.L^{-1}$. This means that an additional downstream step (activated carbon, for example) is needed to eliminate these by-products.

Alachlor is effectively oxidized by ozone in drinking water systems, but numerous by-products have been identified. For example, the various chemical degradation pathways of alachlor include oxidation of the aryl-ethyl group, N-dealkylation, cyclization and cleavage of the benzene ring. Intermediate reactions give rise to numerous by-products that include an aromatic nucleus in their structure.

Ozonation of alachlor leads to the following by-products: N(2,6-diethylphenyl)-methylene amide, 8-ethyl-3,4-dihydroquinoline, 8-ethylquinoline, 1-chloroacetyl-2-hydro-3-ketone-7-acetylindole, 2-chloro-2',6'-diacetyl-N-(methoxymethyl)-acetanilide, 2-chloro-2'-acetyl-6'-ethyl-N-(methoxymethyl)-acetanilide, two hydroxylated alachlor isomers, as well as 1-chloroacetyl-2,3-dihydro-7-ethylindole, 2-chloro-2',6'-ethylacetanilide, 2-chloro-2',6'-acetylacetanilide and 2-chloro-2'-ethyl-6'-acetyl-N-(methoxymethyl)-acetanilide.

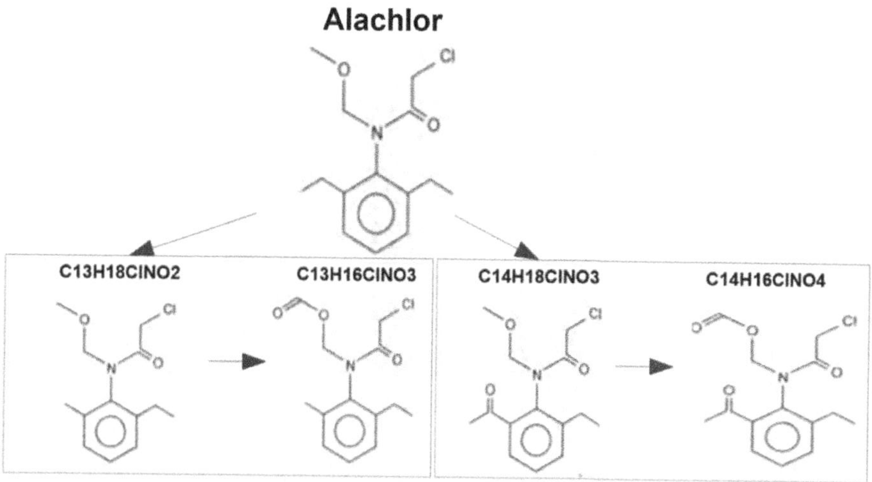

Figure 2.6. *Alachlor oxidation by-products*

2.4.1.1.2. Drug residues

The mechanisms of ozonation of drug residues are similar to those of pesticides, depending on the structure of the molecule, its concentration and the conditions of ozonation. Clearly, however, ozone has great potential for the oxidation of certain pharmaceutical products and EDs in drinking water treatment plants.

An oxidation efficiency of 92% is obtained for diclofenac with an ozone dose of 0.21 g $O_3.g^{-1}$ DOC, and 99% with an ozone dose of 0.41 g $O_3.g^{-1}$ DOC. The latter dose should be injected to achieve over 90% elimination of carbamazepine and atenolol. The elimination kinetics of medroxyprogesterone are a little slower, as only 57% elimination is achieved with a Ct of 1.03 mg.min.L^{-1}, compared with over 90% for norethindrone and levonorgestrel. A Ct of 1.85 mg.min.L^{-1} is required to achieve over 80% elimination of caffeine.

Norethindrone and levonorgestrel contain ethynyl groups that are more reactive with ozone than substances such as caffeine and progesterone.

Substances that react rapidly to O_3, such as clindamycin, erythromycin, mefenamic acid, naproxen, sulfapyridine, trimethoprim and sotalol, are oxidized to over 90% at a dose of 0.21 g O_3.g^{-1} DOC.

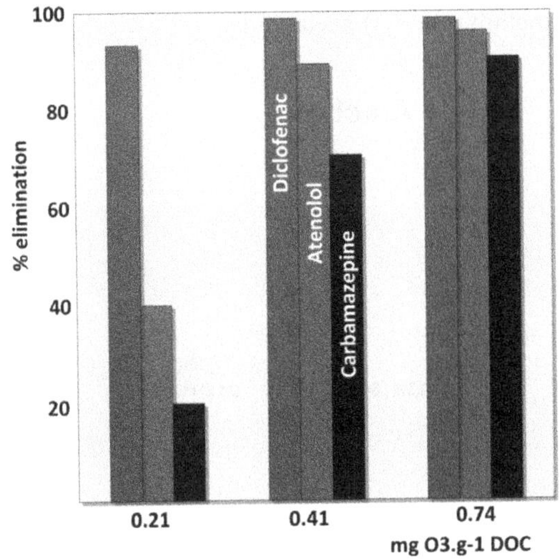

Figure 2.7. *Percentage of drug elimination with different ozone doses (DOC: dissolved organic carbon). For a color version of this figure, see www.iste.co.uk/gaid/reuse2.zip*

Trimethoprim is oxidized with an ozone dose of 0.3 mg.L^{-1}, whereas a dose of 0.41 g O_3.g^{-1} DOC is required for the elimination of gemfibrozil, estrogenic steroids (estrone, 17-β-estradiol, estriol and 17-α-ethinylestradiol).

Slow-reacting substances such as ibuprofen, paracetamol, sulfamethazine, sulfadiazine and phenazone are oxidized to over 90% at a dose of 0.81 g O_3.g^{-1} DOC, whereas bezafibrate requires a dose of at least 0.60 g O_3.g^{-1} DOC.

It thus appears that the ozone doses generally applied in drinking water treatment (2–3 mg O_3.L^{-1}) lead to reductions of over 95% for many drug residues for contact times of 2–3 min in ozonation tanks.

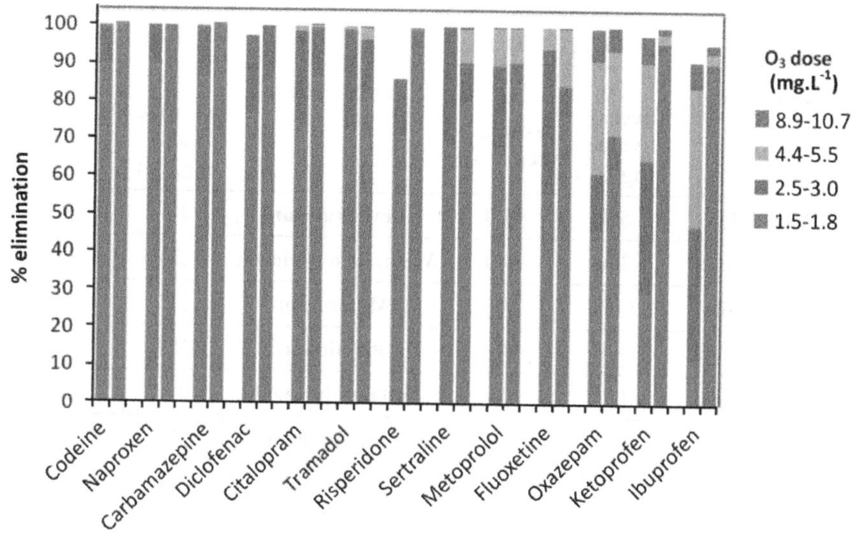

Figure 2.8. *Ozonation performance at pH 6.0 for different O_3 dosages (6.9 mg.L^{-1} DOC). For a color version of this figure, see www.iste.co.uk/gaid/reuse2.zip*

A ratio greater than 1.2 g O_3.g^{-1} COD is recommended to remove more than 80% of micropollutants in water containing 6.9 mg.L^{-1}-DOC. On the other hand, a ratio greater than 1.5 g O_3.g^{-1} DOC gives yields of over 90%, and above all enables refractory micropollutants to be oxidized.

Figure 2.8 shows the importance of the O_3 dose in achieving high efficiency in water loaded with natural organic matter, as the demand for O_3 due to organic matter is high. It may therefore require high dosages even if the concentration of micropollutants to be removed is low. When the water to be treated contains a high concentration of DOC, this leads to an immediate demand for O_3, which can result in a reduction in O_3 performance with regard to micropollutants.

Ozone oxidizes all micropollutants with an electron-rich configuration, but its reactivity also depends on the presence of the natural organic molecules that contribute to ozone demand.

For many drug substances that react well with O_3, an additional reaction with OH· radicals is not necessary, as the reaction time of these substances with ozone is faster than the time required for the formation of free radicals.

Substances	Ozonation (6 mg.L^{-1} O_3)		Substances	Ozonation (6 mg.L^{-1} O_3)	
	Co average value (ng.L^{-1})	% average removal		Co average value (ng.L^{-1})	% average removal
Atenolol	500	95.0	**Levetiracetam**	81	94.0
Enolic acid	380	90.0	**Methylbenzotriazole**	3,267	96.0
Benzotriazole	3,633	77.6	**Metoprolol**	313	97.0
Carbamazepine	390	99.0	**Primidone**	103	90.2
Clarithromycin	153	97.0	**Sulfamethoxazole**	297	98.0
Diclofenac	813	99.0	**Trimethoprim**	29	82.5
Gabapentin	1,017	66.6	**Valsartan**	81	93.7
Hydrochlorothiazide	1,213	99.0	**Venlafaxine**	267	96.0

Co: initial concentration.

Table 2.5. *Ozonation performance on drug residues*

The results obtained clearly illustrate the high efficiency of chemical oxidation with ozone under the following operating conditions: contact time 15 min, O_3 dose 8–10 g.m^{-3}.

With the exception of gabapentin (66.6%), all other molecules are more than 80% removed by ozone, with 99% efficiency for carbamazepine, diclofenac, hydrochlorothiazide and sulfamethoxazole (98%).

Ozonation by-products of drug substances have been identified many times. The ozonation of diclofenac (Figure 2.10) leads to intermediate products such as 2-[(2,6-dichlorophenyl)amino]-5-hydroxyphenylacetic acid (D1) and 2,6-dichloro-4-hydroxybenzenamine (D8).

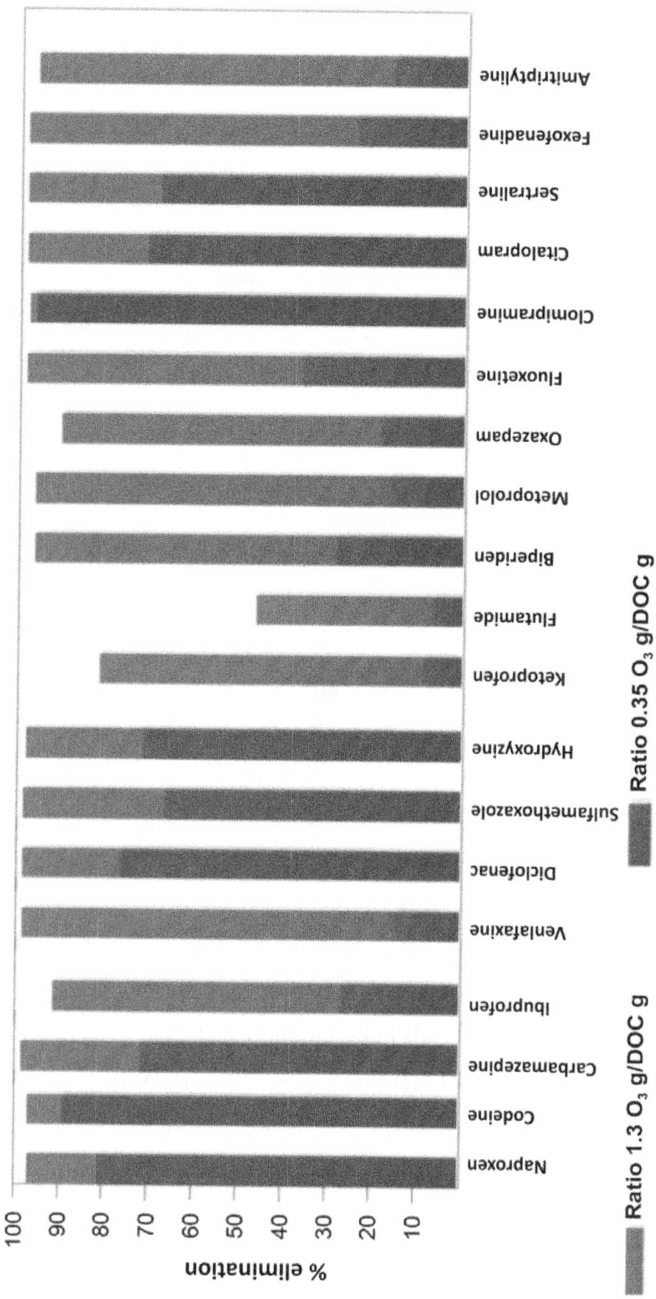

Figure 2.9. *Ozonation of micropollutants at various g O3/g DOC ratios (DOC at inlet 5 mg/L). For a color version of this figure, see www.iste.co.uk/gaid/reuse2.zip*

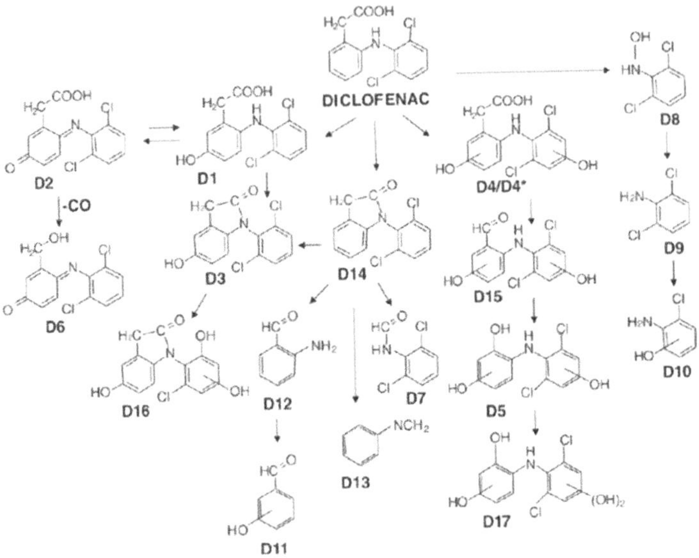

Figure 2.10. *Identification of diclofenac ozonation by-products*

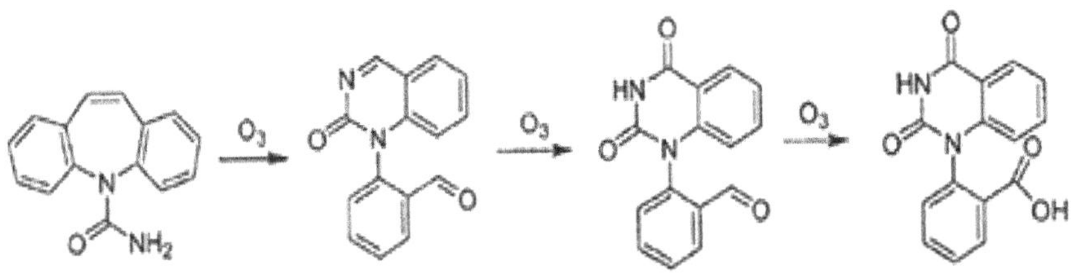

Carbamazepine BQM BQD BaQD

Figure 2.11. *Identification of carbamazepine ozonation by-products*

The ozonation of carbamazepine leads to by-products such as:

– 1-(2-benzaldehyde)-4-hydro-(1H,3H)-quinazoline-2-one (BQM);

– 1-(2-benzaldehyde)-(1H,3H)-quinazoline-2,4-dione (BQD);

– 1-(2-benzoic acid)-(1H,3H)-quinazoline-2,4-dione (BaQD).

2.4.2. *Advanced chemical oxidation*

When ozonation kinetics are slow, it is possible to speed up ozone oxidation reactions by combining them with hydroxyl radicals (OH·). Advanced oxidation processes (AOP) are an effective solution for degrading refractory compounds.

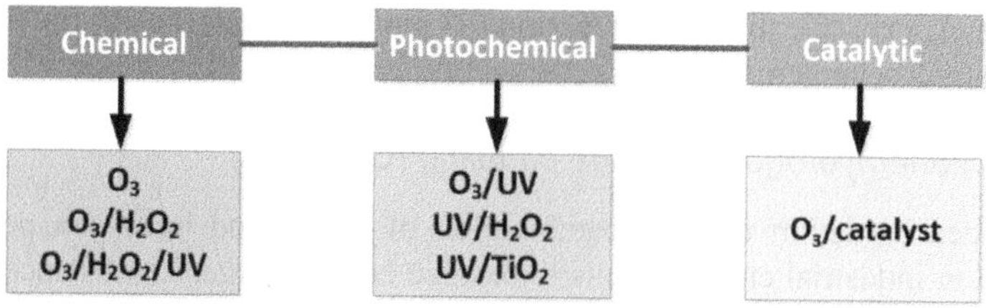

Figure 2.12. *Some advanced oxidation solutions*

The aim of AOPs is to degrade refractory and non-biodegradable compounds into various by-products, and eventually into inert end-products. Reaction mechanisms are based on the formation of free radicals (von Gunten 2003). However, the metabolites formed during these reactions may be harmful compounds. Wherever possible, these intermediates must also be removed by more stringent operating conditions.

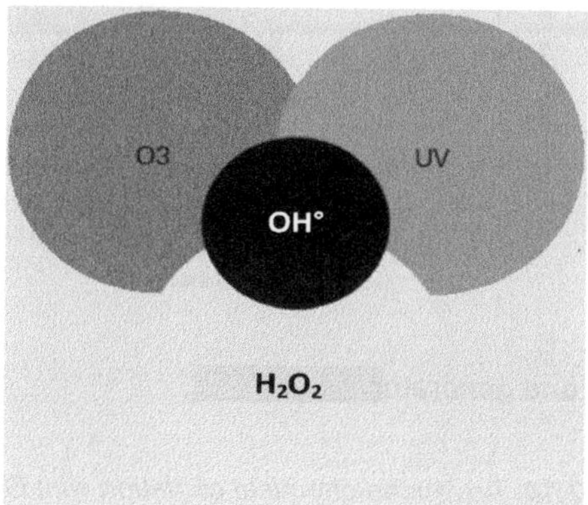

Figure 2.13. *Main combinations used in drinking water*

AOPs are therefore characterized by a variety of free-radical reactions involving combinations of chemical agents such as ozone (O_3), hydrogen peroxide (H_2O_2), transition metals, metal oxides and auxiliary energy sources such as visible ultraviolet (UV) and ultrasound. Examples of AOPs include O_3/H_2O_2, O_3/UV,

$O_3/H_2O_2/UV$, H_2O_2/UV, Fenton (Fe^{2+}/H_2O_2) and heterogeneous photo-oxidation using titanium dioxide (TiO_2). Hydroxyl radicals (OH·) are the main oxidizing agent in AOPs, while other radical species such as superoxide radical anions ($O2·^-$), hydroperoxyl radical (H-O-O·), triplet oxygen ($^3O_2$) and organic peroxyl radical compounds (R-O-O·) are also involved.

In wastewater, the main combinations (Figure 2.13) are based on the use of ozone and hydrogen peroxide.

2.4.2.1. *Ozone/hydrogen peroxide coupling (O_3/H_2O_2)*

Advanced oxidation using a combination of ozone and hydrogen peroxide is often used in industrial environments to oxidize color or organic substances that are difficult to biodegrade. It is also used in wastewater treatment to oxidize micropollutants and refractory DOC. In general, chemical activation is often used to activate the primary oxidant and induce the formation of a secondary oxidant. In most cases, the secondary oxidant is a hydroxyl radical (HO·). The oxidation potentials of molecular ozone and hydroxyl radicals are 2.07 and 2.80 V, respectively, indicating that they are very powerful oxidants. When combined with hydrogen peroxide, ozonation is very beneficial for wastewater reuse.

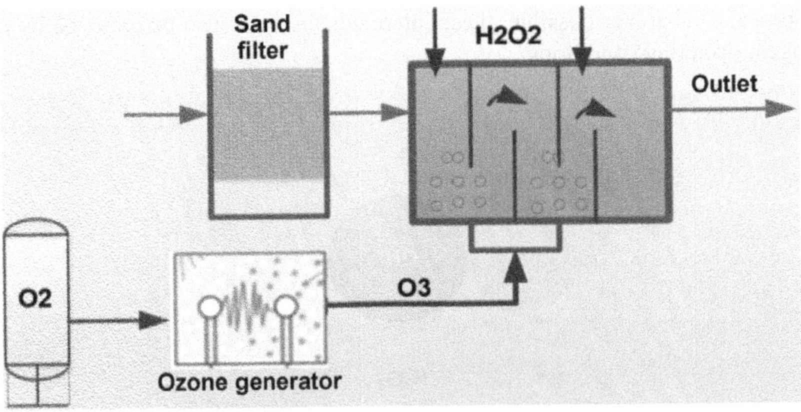

Figure 2.14. *Advanced chemical oxidation with O_3/H_2O_2*

Hydrogen peroxide is injected into the ozonation chambers (Figure 2.14). The addition of H_2O_2 in the presence of ozone reinforces the oxidizing power of ozone toward organic matter (total organic carbon (TOC)) and micropollutants. Hydrogen peroxide (H_2O_2) causes very rapid (instantaneous) decomposition of ozone, giving rise to highly oxidizing and not very selective radical species. Most of these are OH·

hydroxyl radicals, whose redox potential (2.8 V) is much higher than that of ozone (2.07 V).

Oxidation reactions involving hydroxyl radicals in the presence of organic substrates (in aqueous media) are mainly electrophilic addition reactions and hydrogen abstraction reactions. The equations proposed to describe these addition and abstraction reactions are as follows:

– first case: addition of the OH· radical to the organic compound R:

$$R + OH\cdot \rightarrow (R\text{-}OH\cdot) \rightarrow \text{hydroxylated products}$$

– second case: removal of a hydrogen atom:

$$RH_2 + OH\cdot \rightarrow (R\text{-}H\cdot) + H_2O \rightarrow \text{oxidized products}$$

In both cases, organic radicals are formed, which may subsequently react with other radicals (termination reaction) or react with another molecular oxidant in solution (propagation reaction). The initial activation of the molecular oxidant leads to the generation of hydroxyl radicals (initiation reaction).

Combined O_3/H_2O_2 oxidation can degrade all types of micropollutants, except those saturated with chlorine. In contrast to ozonation, which is only effective on a limited number of organic molecules when used as a pure oxidation process, this process is not very selective and has a broad spectrum of action.

Hydrogen peroxide reacts very rapidly in its ionized form (HO_2^-, pKa = 11.6) with ozone to form free radicals (OH·) according to the following reaction:

$$2O_3 + H_2O_2 \rightarrow 2OH\cdot + 3O_2$$

Once released during the reaction of HO· with ozone, these radicals then initiate other radical mechanisms for the decomposition of hydrogen peroxide.

The reaction speed of hydroxyl radicals on organic molecules can be 1,000 to 1,000,000 times greater than that obtained with molecular ozone. As a result, the contact times required for combined O_3/H_2O_2 oxidation are shorter.

Process efficiency depends on ozonation rate, contact time, injected H_2O_2/injected O_3 ratio, TAC, pH, TOC and temperature.

It is preferable to remove this H_2O_2 residual before final disinfection with chlorine or ClO_2, as H_2O_2 consumes Cl_2 and ClO_2, resulting in unnecessary overconsumption of Cl_2 and ClO_2.

The consumption ratio is theoretically equal to 2.09 mg Cl_2/mg H_2O_2, but in practice, ratios of 3.00 mg Cl_2/mg H_2O_2 are obtained, as there is a demand for chlorine from organic matter or other mineral elements.

2.4.2.2. *UV/H_2O_2 coupling*

Ultraviolet (UV) radiation is widely used as a wastewater disinfection process (see Chapter 8 of Volume 1). When applied at high doses (1,000–2,000 $mW.s^{-2}$), UV radiation breaks down organic compounds, including pharmaceutical products, by direct photolysis (phenazone, ketoprofen, diclofenac, sulfamethoxazole, sulfamonomethoxine and antipyrine), or by indirect photolysis through coupling with another oxidant. Indeed, hydroxyl radicals are generated in the UV/H_2O_2 process and promote degradation mechanisms.

Degradation by UV photolysis depends on UV photon absorption and quantum efficiency. Consequently, UV photolysis is also a selective process. In addition to the radical oxidation mechanism provided by H_2O_2, UV radiation itself can also degrade compounds by direct photolysis. The efficiency of this process depends on the spectrum of the UV lamp used, the absorption spectra of the water to be treated and the absorption spectra of the target compounds. Operating parameters include UV dose, H_2O_2 dose and water pH.

Pesticides such as metaldehyde do not absorb UV light and are not degraded, while atrazine, with its high molar absorption coefficient (3,683 $m^{-1}cm^{-1}$) and high quantum efficiency, shows significant conversion.

Reaction rate constants for hydroxyl radicals produced by O_3/H_2O_2 and UV/H_2O_2 range from 10^6 and 10^9 $M^{-1}s^{-1}$. For example, the reaction rate constant for atrazine is 2.7×10^9 $M^{-1}s^{-1}$.

Hormones are well degraded with a conventional UV dose (300 $mJ.cm^{-2}$) in the presence of H_2O_2. The most commonly used lamps are medium-pressure (MP) and low-pressure (LP) mercury vapor lamps. MP lamps emit a broad spectrum of light, between 200 and 800 nm, with a relevant range between 235 and 300 nm for UV/H_2O_2 coupling. This wavelength range corresponds to the absorption spectrum of many compounds, resulting in high photolytic capacity, but also lower efficiency and higher potential for by-product formation.

Low-pressure lamps emit UV light at a single wavelength (253.7 nm). Consequently, they can only photolyze compounds with an absorbance around 254 nm.

Oxidation results with MP (2,200 W) and LP (300 W) lamps and 5 mg.L^{-1} H_2O_2 are summarized in Figure 2.15 for pesticides and drugs. Water transmittance is 80%.

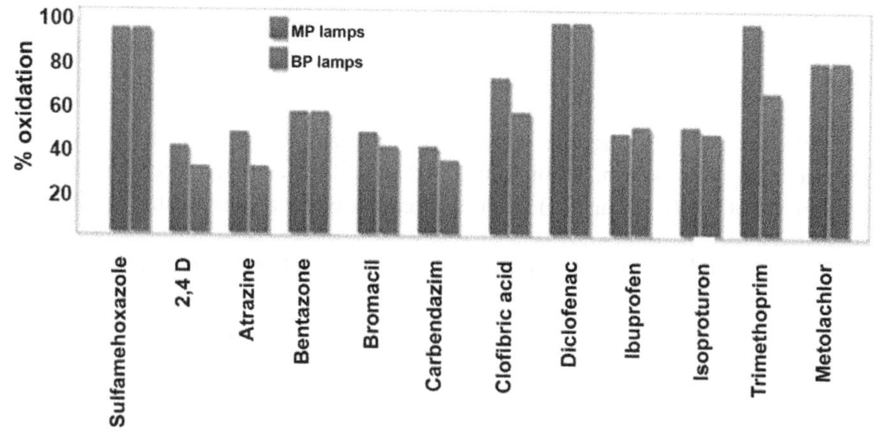

Figure 2.15. *Percentage of pesticide and drug oxidation with UV/H_2O_2. For a color version of this figure, see www.iste.co.uk/gaid/reuse2.zip*

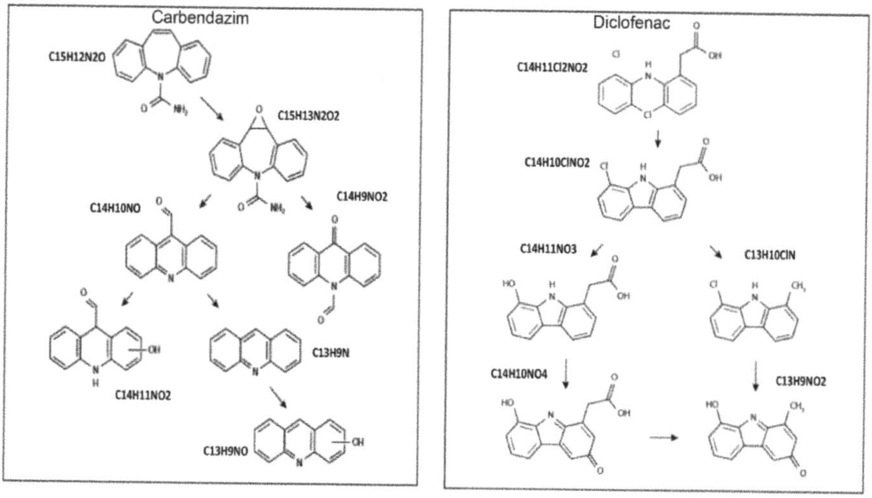

Figure 2.16. *By-products identified with UV/H_2O_2 for carbamazepine and diclofenac*

UV/H_2O_2 treatment using a dose of 300 mJ.cm^{-2}, combined with an injection of 5–10 mg.L^{-1} of H_2O_2, reveals a difference between the efficiency of LP and MP lamps for some compounds (2,4 D and trimethoprim), while for other compounds (sulfamethoxazole, carbendazim, diclofenac, metolachlor and isoproturon), no difference is observed. There are various oxidation by-products, as shown in Figure 2.16 for carbamazepine and diclofenac.

Pesticides such as atrazine and its derivatives, or pharmaceuticals such as clofibric acid, amoxicillin, ampicillin and cloxacillin, are well decomposed after addition of H_2O_2 (10–50 mg.L^{-1}) in the UV reactor under a contact time of 30 min.

Due to the high UV doses required, this process is not economically viable.

Although AOPs are recognized as an effective technique for oxidizing and mineralizing almost all organic contaminants, high energy consumption is one of the disadvantages holding back the widespread use of these techniques.

2.4.3. *Activated carbon adsorption*

Activated carbon has long been known for its ability to adsorb organic substances in water, as well as the molecules responsible for tastes and odors.

It was therefore a natural choice for removing organic micropollutants (EDs, drug residues and industrial residues) in the same way as pesticides.

In wastewater treatment, activated carbon adsorption is currently a widely used technique, as the transfer of pollution onto this porous medium is currently the most reliable process, provided it is properly sized and operated.

Activated carbon technologies depend on the type of activated carbon used. They are designed either:

– with PAC or micro-grains activated carbon in specific reactors, operating with recirculation or by a one-off addition in a decanter;

– GAC in a filter.

Activated carbon is available in different particle sizes: PAC with diameters between 5 and 35 μm, micro-grain activated carbon (μgrains) with diameters between 0.2 and 0.6 mm, and GAC with average diameters between 0.85 and 1.2 mm.

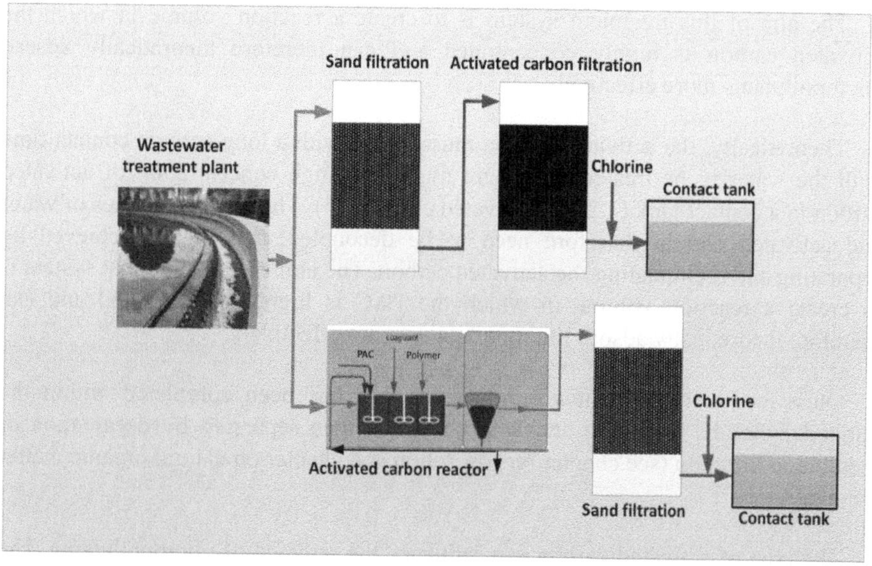

Figure 2.17. *Disposal of activated carbon in the water reuse process*

2.4.3.1. *Powdered activated carbon and micro-grain activated carbon*

Activated carbon is best used in powder or micro-grain form, as it enables dosages to be combined and controlled according to the pollution entering the system. It is therefore the product water quality/operating cost pairing that is often chosen by engineering firms and local authorities, as it enables them to optimize the operating cost of activated carbon throughout the year according to seasonal variations in pollution.

To this end, water treatment companies have brought various processes to market, based on PAC reactors. Today, design offices tend to formalize this type of process in their specifications, whether for tertiary or quaternary treatment of wastewater, or for refining treatment in drinking water plants.

PAC or micro-grain activated carbon reactors are used for natural organic matter, pesticides and micropollutants. All the more so as all these substances can end up in surface waters simultaneously. These reactors are being developed in Europe, the United States and China. They are applied to both drinking water and wastewater. In the case of wastewater, they are installed for tertiary treatment, after clarification.

Micropollutant removal efficiency is based on the adsorption capacity of activated carbon.

The aim of this treatment system is to create a reaction volume in which the activated carbon is highly concentrated and can therefore theoretically adsorb micropollutants more effectively.

Theoretically, the activated carbon must be left with a long enough contact time with the water to be treated, and there must be a high concentration of activated carbon in a contact tank (1.2–5 g activated carbon.L^{-1}). The residence times of water and activated carbon therefore need to be decoupled, and this is achieved by separating and recirculating the activated carbon. The aim of this treatment system is to create a reaction volume in which the PAC is highly concentrated and can therefore theoretically adsorb micropollutants more effectively.

Once the micropollutant adsorption reaction has been completed within the allotted hydraulic time, the activated carbon is then separated by decantation or membrane filtration (see chapter on adsorption and chapter on natural organic matter removal).

The type of activated carbon can influence the removal of micropollutants. The selection of activated carbons is based on their technical characteristics (pore size, specific surface, iodine value, etc.), as well as on the results of on-site laboratory or pilot tests.

Adsorption with PAC or micro-grain activated carbon is governed by three parameters:

– PAC dosage;

– the hydraulic residence time in the activated carbon reactor, so as to ensure that the adsorption balance between activated carbon and organic micropollutants is reached;

– the age of the activated carbon in the AC reactor, to fully utilize the carbon's adsorption capacity (the carbon is separated after passing through the reactor, and then recirculated upstream of the reactor).

The general adsorption mechanisms for pesticides and other micropollutants are similar to those described for natural organic matter, but they differ in that micropollutants are much smaller in size; it is the mesopores and micropores that are essential.

PAC has a higher adsorption capacity and adsorption rate than micro-grain carbon and GAC. This is all the more important as the removal efficiency of micropollutants is based on the adsorption capacity of activated carbon.

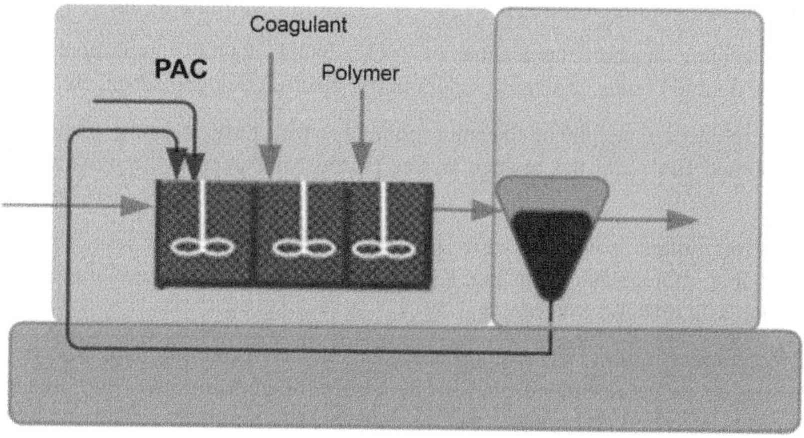

Figure 2.18. *PAC reactor with recirculation*

Adsorption is a surface phenomenon whereby molecules bind to the surface of activated carbon using a variety of forces (electrical charges, dipole/dipole interactions, Van der Waals forces, covalent bonds, hydrogen bonds, etc.). Hydrophobic interaction processes on the carbon surface are the leading mechanisms for removing micropollutants. The adsorbed concentration of a substance at equilibrium can be calculated empirically by the Freundlich isotherm (most often used to describe adsorption):

$$\frac{(Cinlet - Coutlet)V}{M_{AC}} = k\ Cs^1/n$$

with:

– Cinlet: concentration of substance in raw water (ng or $\mu g.L^{-1}$);

– Coutlet: concentration of substance in treated water (ng or $\mu g.L^{-1}$);

– M_{AC}: mass of PAC or micro-grains in the reactor (g);

– V: activated carbon reactor volume (L);

– k: Freundlich adsorption constant related to adsorption capacity (μg micropollutant.g^{-1} AC) ($L^{1/n}.\mu g^{-1/n}$ micropollutant);

– 1/n: Freundlich parameter, linked to adsorption intensity and reflecting the degree to which adsorption is a function of contaminant concentration.

The parameters that influence the adsorption of micropollutants are:

– molecular structure: presence of NO_2, NOH, COOR functional groups, halogens, double bonds, coplanarity, symmetry, electron delocalization, etc.;

– the molecular weight of the micropollutants (the lightest being adsorbed, but also desorbed first) and the molecular size (it must not exceed the maximum pore size);

– micropollutant solubility: the higher the solubility, the less adsorption. Hydrophobic molecules (high log Kow) are better adsorbed, and highly polar compounds are difficult to remove;

– the concentration of the micropollutant: equilibrium depends on the concentration of the dissolved phase. The higher the concentration, the greater the proportion adsorbed (although the percentage of removal remains relatively constant);

– competition between molecules. In the case of mixtures, the most difficult substances to adsorb are even less adsorbed than when tested separately. The presence of small-scale natural organic matter (comparable to micropollutants) in water reduces the adsorption efficiency of micropollutants, as it competes for binding sites or can block the pores of the structure;

– temperature: increasing temperature shifts the sorption equilibrium and reduces micropollutant adsorption.

2.4.3.2. *Sizing parameters*

2.4.3.2.1. Choice of activated carbon

The choice of activated carbon is important, as water contains high molecular weight molecules (1,000 Da) and micropollutants (50–200 Da).

It is important to define the right quantity of carbon and to choose the carbon with the appropriate specificities for the characteristics of the water to be treated, as the adsorption of organic molecules is based on mechanisms that depend on physical processes such as electrostatic attractions (dipole/dipole interactions, Van der Waals-type interactions) or chemisorption processes.

The adsorption phenomenon is only significant if the adsorbent medium has a well-developed pore structure and, consequently, a large specific surface area.

However, the effectiveness of an activated carbon, with regard to micropollutants, is linked to its adsorption capacity, with regard to these molecules, and to their diffusion kinetics in the pores. Adsorption capacity depends on temperature and pressure, the concentration of organic molecules, functional groups on the surface, the presence of other adsorptive constituents such as DOC and competition between different molecules.

2.4.3.2.2. Sizing parameters based on isotherms

The parameters obtained from adsorption isotherms provide information on the adsorption mechanism, surface properties and affinity between the adsorbate and the adsorbent.

The empirical Freundlich model frequently used for solid adsorbents takes into account the possible formation of several successive layers of molecules on the surface of the solid, implying heterogeneity in the energy of the adsorption sites on the adsorbent. In this case, the isotherm is described by the following equation:

$$qe = k\,Ce^{1}/n$$

where q_e represents the adsorption capacity (mg adsorbate/g adsorbent), C_e is the concentration of the pollutant at equilibrium (mg.L^{-1}), k is the constant describing the adsorption capacity of the activated carbon (mg.g^{-1}) (L$^{1/n}$.mg$^{-1/n}$) and $1/n$ represents a second constant linked to the heterogeneity of the activated carbon surface, describing the affinity between the adsorbent and the pollutant.

Also with:

$$qe = V\,(Co - Ce)/M$$

where V is the volume of water containing the pesticides (L), C_o and C_e are, respectively, the initial and equilibrium concentrations of the organic substances (mg.L^{-1}) and M is the mass of activated carbon (g).

The Freundlich equation leads to the following linear form:

$$\log qe = 1/n\,\log Ce + \log \mathrm{k}$$

By plotting log q_e as a function of log (C final) for each PAC concentration studied, it is possible to determine $1/n$ from the slope of the line and log k, being the y-intercept of the origin.

The value of $1/n$ lies between 0 and 1. The more heterogeneous the carbon surface, the closer the value of $1/n$ is to 0.

According to Langmuir, saturation of adsorption sites occurs in a single layer. These sites can contain only one molecule per site. There are no interactions between the adsorbed molecules. In this case, the adsorption sites are considered to be equivalent, so adsorption is homogeneous. Langmuir's model is given by the following equation:

$$qe = \frac{qmaxbCe}{1+bCe}$$

where:

– q_e is the quantity adsorbed per mass of adsorbent in mg.kg^{-1} of activated carbon;

– C_e is the equilibrium concentration in mg.L^{-1};

– b is a constant describing adsorption energy, the Langmuir constant in L.mg^{-1}. It depends on temperature and gives an indication of the affinity of the adsorbate for the adsorbent: the higher it is, the stronger the affinity;

– q_{max} is the maximum quantity that can be adsorbed on a given mass of solid in monolayer adsorption. It is also the number of active sites per unit mass of the solid phase, indicating total occupation of adsorption sites (mg.kg^{-1}).

Plotting $1/q_e$ as a function of $1/C_e$ gives a straight line, whose slope leads to $1/q_{max}.b$ and whose y-intercept gives $1/q_{max}$.

Predicting the simultaneous adsorption of several constituents is one of the most complex problems in the field of adsorption. This is particularly true of surface waters, where the water matrices to be treated are complex and comprise a mixture of pesticides, micropollutants and humic substances.

In general, it appears that the Freundlich equation is more representative of adsorption mechanisms than the Langmuir equation, as the principle of stacking adsorbed organic substances seems more realistic. For this reason, it is accepted that there are few restrictions on the use of the Freundlich model.

2.4.3.2.3. Application parameters

Table 2.6 shows the parameters for applying AC in reactors.

The total quantity of sludge (kg) present in the reactor is the sum of AC, metal hydroxides and TSS. Thus:

$$\text{Volume (m}^3\text{) [AC (kg.m}^{-3}\text{) + TSS (kg.m}^{-3}\text{) + Me(OH)}_3\text{ (kg.m}^{-3}\text{)]}$$

Parameters	Values
AC concentration (mg.L^{-1})	10–25
Hydraulic residence time (min)	15–60
MES inlet concentration (mg.L^{-1})	5–25
Coagulant dosage (mg metal.L^{-1})	1–3
AC concentration in reactor (g.L^{-1})	1–20
Age of AC (days)	6–15

Table 2.6. *AC reactor design parameters (Veolia)*

The applied load (kg.day^{-1}) is given by:

– flow rate (m^3.day^{-1}) [AC (kg.m^{-3}) + TSS (kg.m^{-3}) + $Me(OH)_3$ (kg.m^{-3})];

– AC age: volume (m^3) . conc. AC (kg.m^{-3})/Q-AC extracted (kg.day^{-1});

– AC hydraulic residence time (min) = volume (m^3)/inlet flow (m^3.min^{-1}).

2.4.3.2.4. Operating parameters and performance

The parameters to be optimized are:

– carbon residence time in the contact reactor (recirculation rate, reactor volume);

– AC dosage;

– coagulant and polymer dosage;

– carbon age and frequency of spent carbon extractions.

2.4.3.2.5. Performance of AC reactors in relation to pesticides and other micropollutants

The log Kow (logarithm of the octanol/water partition coefficient) can be a first indicator of the adsorbability of organic compounds such as pesticides and other micropollutants. Kow represents the ratio of a substance's concentration in octanol to its concentration in water. It indicates whether a molecule is hydrophilic (low log Kow) or hydrophobic (high log Kow). It can also be used to estimate the sorption of an organic substance on a soil. Indeed, a high log Kow would indicate a tendency to adsorb to particles. Hydrophobic compounds are better adsorbed than hydrophilic ones. It is also used in conjunction with activated carbon as an adsorption indicator.

2.4.3.3. *Pesticides*

Well-adsorbed pesticides		Moderately adsorbed pesticides	Pesticides with little or no adsorption
Acetochlor	Ethylbenzene	Hydroxyalachlor	Aminotriazole
Acetochlor OXA	Fenoprop	Aldicarb	Glyphosate
Alachlor	Flusilazole	Bentazone	AMPA
Aldrin	Heptachlor	Bromacil	Mesotrione
Ametryne	Heptachlor epoxide	Carbofuran	Nicosulfuron
Atrazine	Isoproturon	Deethylhydroxyatrazine	Sulcotrione
Chlordane	Ioxynil	Imidachloprid	
Cyanazine	Linuron	Iodosulfuron	
Carbendazim	Lindane	Metaldehyde	
Chlorfenvinphos	Malathion	Metolachlor ESA	
Chloridazone	Metolachlor	Metolachlor OXA	
Chlortoluron	Hydroxymetolachlor	Metribuzin	
Chlorpyrifos ethyl	Metazachlor	Metamitron	
Deisopropylatrazine	Parathion	Metsulfuron	
Desethylatrazine	Parathion methyl		
Carbofuran	Pendimethalin		
2,4 D	Pentachlorobenzene		
DDT	Pentachlorophenol		
Dianizon	Phenmedipham		
Dicamba	Picloram		
Dichloprop	Propachlor		
Dichlorvos	Prosulfocarb		
Dimethachlor	Propazine		
Dimethenamid	Simazine		
Dinoseb	2,4,5 T		
Dieldrin	Terbuthylazine		
Diuron	Terbutryne		
Diflufenicanil	Thiodicarb		
Endrin	Triadimenol		
Endosulfan alpha	Triclopyr		
Ethofumesate	Trifluralin		

Table 2.7. *Adsorption of pesticides on activated carbon*

The first sizing parameters are the isotherm constants, which predict the efficiency of the activated carbon in relation to the pesticide concerned.

The tests must be carried out using water from the resource, as the AC doses obtained in synthetic water do not consider the fact that DOC is in competition when it is present in this water. On the other hand, tests are more representative for groundwater, as DOC is present in low concentrations (DOC < 1 $mg.L^{-1}$). In general, pesticides with Freundlich coefficients above 150 ($mg.g^{-1}$) ($L^{1/n}.mg^{-1/n}$) are well eliminated.

Pesticide	PM	log Kow	log Koc	Solubility (ppm)	Freundlich parameters k 1/n ($mg.g^{-1}$) ($L^{1/n}.mg^{-1/n}$)	
Acetochlor	269.11	3.17	2.49	282	350	0.48
Acetochlor ESA	315.11	0.32	2.49	282	132	0.51
Acetochlor OXA	265.13	2.28	1.38	–	298	0.47
Alachlor	269.77	3.52	2.52	242	400	0.52
Alachlor OXA	265	–	–	–	–	–
Hydroxyalachlor	285.77	2.74	–	–	127	0.43
Deschloroalachlor	–	3.08	–	–	331	0.46
Aldicarb	190.26	1.13	1.47	6,000	102	0.47
Aldrin	364.91	3.01	–	0.027	651	0.92
Ametryne	227	2.98	–	209	201	0.52
Aminotriazole	84.08	–0.97	–	–	100	0.65
AMPA	186.17	–2.17	–	–	–	–
Atrazine	215.69	2.68	1.94	33	179	0.48
Bentazone	240.3	–0.46	1.32	500	102	0.62
Bromacil	261.12	2.11	1.86	815	44	0.47
Carbofuran	221.6	1.98	1.46	700	51.1	0.28
Chlordane	409.78	3.32	5.14	600	320	0.55
Cyanazine	240.7	2.04	1.74	171	208	0.71
Carbendazim	191.19	1.52	3.44	29	190	0.57
Chlorfenvinphos	359.57	4.22	–	145	452	0.58
Chloridazone	221.65	1.50	2.08	–	152	0.48
Chlorsulfuron	357.8	–0.99	1.40	12,500	–	–
Chlortoluron	212.68	2.41	2.29	74	285	0.41
Chlorpyrifos ethyl	350.58	4.23	3.86	–	425	0.61
Desethylatrazine	187.60	1.5	1.38–3.47	3,200	131	0.41

Deisopropylatrazine	173.60	1.2	–	670	123	0.54
Deethylhydroxyatrazine	169	0.43	–	–	59	0.71
2,4 D	221	2.81	1.17	680	125	0.47
DDT	354.5	5.7–6.90	5.17	0.003	322	0.35
Dianizon	304.35	3.81	3.17	60	191	0.28
Dicamba	221.04	2.21	0.30	4,500	131	0.47
Dichloprop	235.07	2.29	3	350	139	0.55
Dichlorvos	221	1.90	1.47	18,000	290	0.45
Dimethachlor	255.18	2.17	1.83	2,300	229	0.50
Dimethenamid	275	2.47	–	1,450	238	0.43
Dinoseb	240	3.15	1.47	52	205	0.44
Diquat	184.24	–4.6	4.5–6.9	712,000	–	–
Diazinon	304.34	3.01	2.72	40	220	0.59
Dieldrine	380.91	5.40	–	0.25	606	0.52
Diflufenicanil	394.3	4.20	–	0.05	400	0.36
Diuron	233.1	2.81	2.68	42	128	0.44
Endrin	380.91	3.20	4	0.23	304	0.42
Ethylene dibromide	187.85	1.76	1.35	4,300	666	0.80
alpha-Endosulfan	406.92	–	2.47–3.30	0.06	194	0.51
Ethofumesate	286.34	2.7	2.50	50	185	0.63
Ethylbenzene	106.17	2.53	–	–	178	0.57
Fenoprop	269.5	3.80	2.47	140	295	0.39
Flusilazole	315.40	3.74	3.22	41.9	265	0.50
Glyphosate	169.07	–3.20	–	–	–	–
Heptachlor	373.32	4.41	4.71	0.06	530	0.65
Heptachlor epoxide	389.3	2.7	2.34	0.35	1,038	0.70
Imidacloprid	255.05	0.57	2.32	610	143	0.76
Iodosulfuron	529.24	1.59	–	25,000	112	0.64
Ioxynil	370.92	2.20	–	3,034	135	0.48
Isoproturon	206.29	2.5	2.24	67.6	233	0.43
Lindane	290.83	–	3.10	7.3	256	0.49
Linuron	249.1	2.19	2.93	75	537	0.38
Mecoprop	214	–0.19	1.30	250,000	–	–
Metaldehyde	176	0.12	2.38	188	95	0.45

Methoxychlor	345.65	3.62	5.90	0.1	212	0.36
Metolachlor	283.8	3.13	2.35	530	559	0.52
Metolachlor ESA	329.11	0.32	1.20	212,461	142	0.48
Metolachlor OXA	279.14	<1	0.9	8,500	142	0.48
Hydroxymetolachlor	300.8	2.81	–	–	289	0.44
Metribuzin	214.29	1.6	1.27	1,100	98	0.60
Malathion	330.35	2.70	–	148	296	0.45
Mecroprop	215	–0.19	–	620	–	–
Mesotrione	283.8	<1	–	160	86	0.68
Metamitrone	202.22	0.85	1.95–2.60	1,700	32	0.19
Metazachlor	277.76	2.5	2.34	450	400	0.48
Metsulfuron	381	<1	1.54	2,800	78	0.52
Nicosulfuron	410.1	0.61	1.47	7,500	35	0.50
Pendimethalin	281.31	5.20	4.55	0.38	500	0.45
Parathion methyl	263.20	2.9	3.7	37.7	265	0.56
Pentachlorobenzene	249.5	5.30	–	–	500	0.45
Pentachlorophenol	270.37	5.66	4.62	–	451	0.42
Phenmedipham	300.31	3.59	2.94	1.80	405	0.61
Picloram	241.46	1.9–2.30	1.20	4.30	185	0.52
Propachlor	211.69	4.48	1.90	–	352	0.43
Prosulfocarb	251.39	4.41	3.22	13.2	285	0.51
Quinmerac	221.5	–1.41	1.93	107,000	–	–
Simazine	201.66	2.51	2.09	3.5	387	0.25
Sulcotrione	–	–1.70	0.03–0.95	1,670	12	0.12
2,4,5 T	255.48	3	4.90	238	358	0.52
Terbuthylazine	229.72	3.21	0.53	6.6	314	0.42
Terbutryne	241.36	3.6	3.30	22	412	0.58
Thiodicarb	354.46	2.5	2.54	25–30	268	0.61
Triadimenol	295.77	3.7	2.47	72	445	0.49
Triclopyr	256.47	2.53	2.90	210–440	183	0.53
Trifluralin	335.28	5.07	3.90	0.3	425	0.50

Table 2.8. *K and 1/n parameters of the Freundlich model*

Table 2.8 shows the K and $1/n$ values for a number of pesticides, enabling us to predict the PAC doses that need to be injected into the reactors to achieve a residual concentration less than or equal to 0.1 $\mu g.L^{-1}$. In this respect, it is interesting to note that the more soluble the compound, the less adsorbable it is, and that this is reflected in the values of the Kow partition coefficient (Figure 2.19).

Glyphosate and AMPA adsorb very little, if at all, to activated carbon. However, they are well eliminated by clarification (40%), ozonation (>90%) and chlorination (>90%). They are also fairly well eliminated (40–50%) in PAC reactors in which a coagulant is injected. The coagulant's effect is what enables this removal.

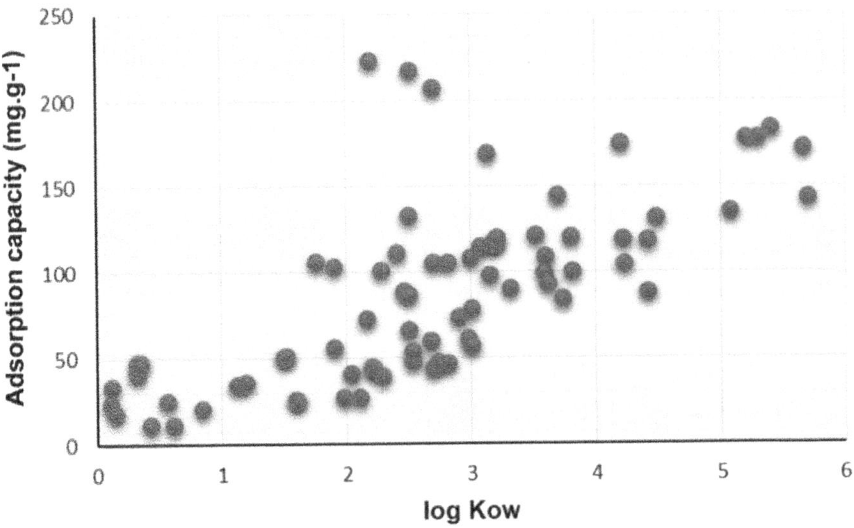

Figure 2.19. *Adsorption capacity ($mg.g^{-1}$) of pesticides as a function of log Kow*

2.4.3.3.1. Efficiency of micro-grain activated carbon

An alternative to decanter reactors is to use PAC in a fluidized bed with micro-grain activated carbon.

The characteristics and adsorption parameters of the activated carbon tested are summarized in Table 2.9.

Micro-grain activated carbon dosages are higher than those obtained with powdered carbon, due to the larger particle size, which reduces exchange surfaces and intrapore diffusion.

Parameters	Activated carbon micrograins
	Microsorb 400 R chemviron
Origin	Coal
d_{10} (μm)	250
d_{50} (μm)	500
Specific surface area ($m^2.g^{-1}$)	860
Apparent density ($kg.m^{-3}$)	450
Iodine index	800

Table 2.9. *Characteristics and adsorption parameters of micro-grain activated carbon*

However, processes using micro-grain activated carbon have adsorbent concentrations in the reactor of between 200 and 350 $g.L^{-1}$. As a result, pesticide removal kinetics remain significant, even for activated carbon dosages varying between 15 and 30 $g.m^{-3}$.

2.4.3.3.2. Drugs

Table 2.10 summarizes the characteristics of a number of drug residues found in water, which can be used to predict their adsorption to activated carbon. The parameters considered include molecular weight, log Kow, solubility and the presence of COOH, NH_2 and SO_2 functional groups.

As illustrated by the q_e ($mg.g^{-1}$) versus log Kow curve for pesticides, the same approach for drug residues leads to the prediction that negative log Kow values are considered to have a negative impact, while very high values have a positive impact on adsorption. Similarly, low molecular weights (approximately < 200 $g.mol^{-1}$) were considered negative and high molecular weights (approximately > 700 $g.mol^{-1}$) were considered positive. Lastly, despite the fact that micropollutants are by definition trace elements, very high concentrations were considered positive.

Gabapentin and metformin have the lowest molecular weights and negative log Kow, and also contain COOH and NH_2 groups, which reduce adsorption. They should therefore be less adsorbed than other micropollutants. Diclofenac, fenofibrate, estradiol, nonylphenol, triclosan, triclocarban and bezafibrate have very high log Kow values (> 4) and should have a high affinity for activated carbon. Triclosan has a high log Kow (4.76) with a symmetrical, coplanar structure and should adsorb well to activated carbon. Iohexol has a negative log Kow (–3.05), but a very high molecular weight, so should not adsorb well, even if its solubility is not very high.

Micropollutants and pharmaceutical residues, endocrine disruptors	PM	log Kow		Solubility (ppm)	Freundlich parameters k 1/n (mg.g^{-1}) (L.µg^{-1})$^{1/n}$	
Atenolol	266.3	0.15	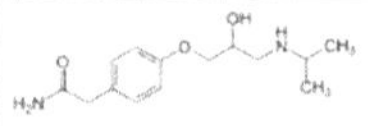	13,330	13	0.22
Bisphenol A	228.3	3.32	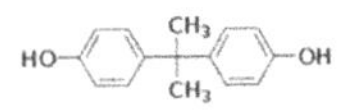	120	35.2	0.23
Bezafibrate	361.8	4.25	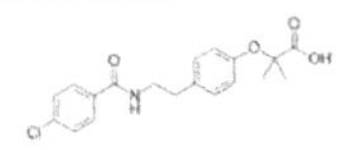	1	5.8	0.24
Carbamazepine	236.2	2.45	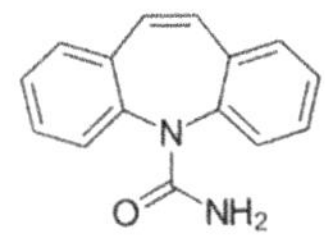	18	9.6	0.21
Caffeine	194.2	–0.07	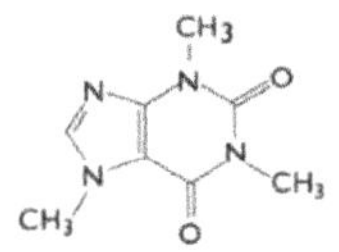	21,600	10.1	0.21
Ciprofloxacin	331.3	0.38	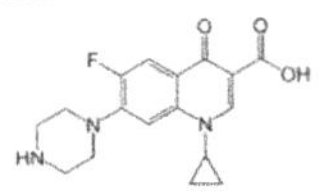	30,000	12.5	0.23
Clarithromycin	748	3.16	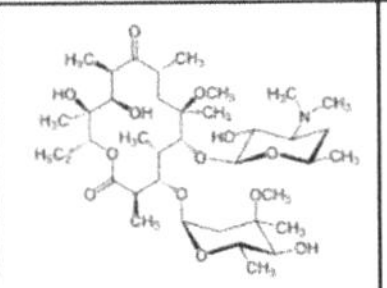	1,693	30.8	0.22
Diclofenac	296.1	4.75	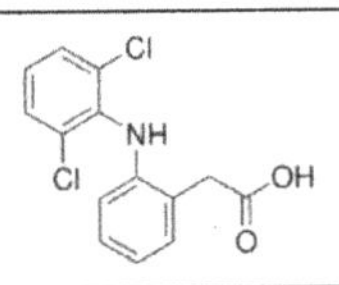	2.37	25	0.32
DEET	191.2	2.40	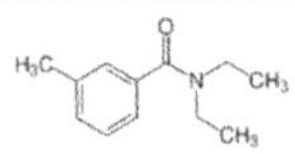	912	8	0.28
Erythromycin	733.9	3.06	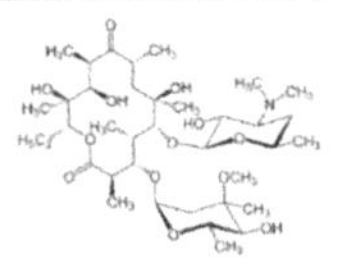	2,000	33.1	0.26

Estradiol	272.4	4.01	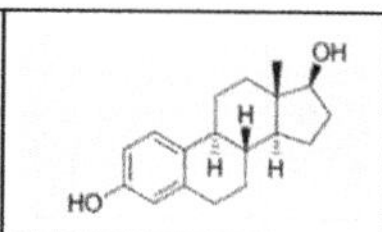	3.6	21.1	0.30
17-α-Estradiol	272.4	4.01	–	0.39	20.5	0.28
Estriol	288.4	2.45	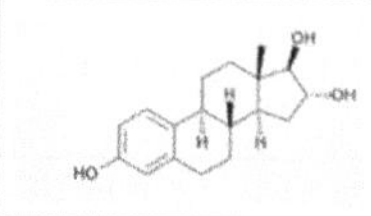	27.4	14.1	0.22
Ethinylestradiol	296.4	3.67	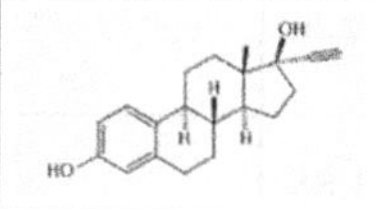	11.3	38	0.25
Fluoxetine	309.3	4.05	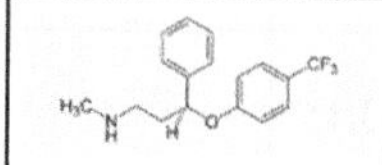	0.017	9.1	0.28
Fenofibrate	360.8	5.20	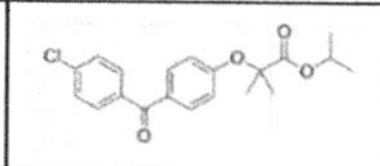	0.42	24.5	0.31
Fluvoxamine	318.3	2.80	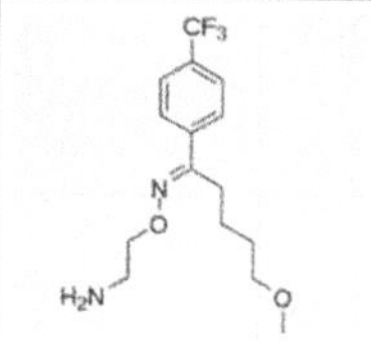	7.34	14.9	0.23
Furosemide	330.7	2.03	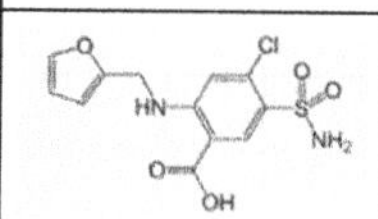	73.1	12.8	0.21
Gabapentin	171.2	–1.10	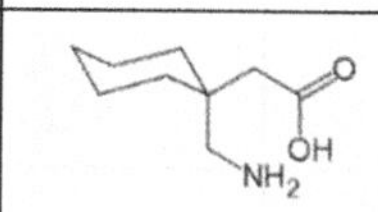	4,490	1.4	0.28
Gemfibrozil	250.3	3.4	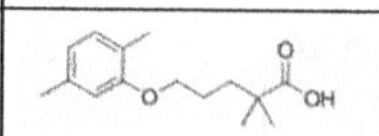	27.8	13	0.22
Hydrochlorothiazide	297.7	–0.07	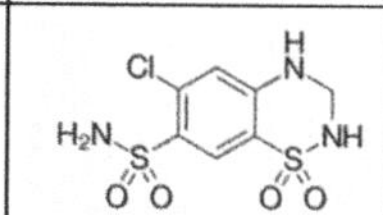	722	10.8	0.28
Ibuprofen	206.4	3.97	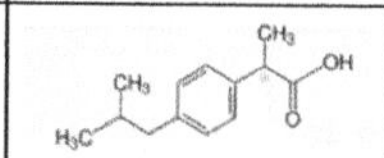	21	33.1	0.27
Iohexol	821.1	–3.05	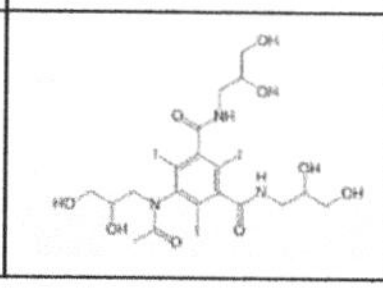	–	2.2	0.24

Iopromide	791.1	2.49	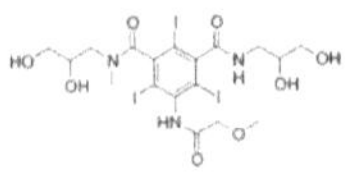	–	4.05	0.28
Iopamidol	777.1	–2.42	–	120,000	22	0.34
Ketoprofen	254.3	3.12	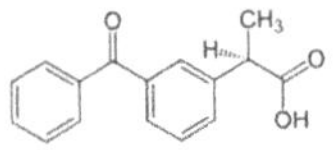	51	35	0.26
Metformin	129.1	–2.60	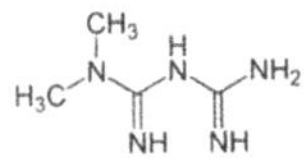	10^6	–	–
Metoprolol	267.3	1.88	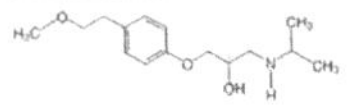	16,041	12.4	0.205
Naxoprene	230.3	3.18	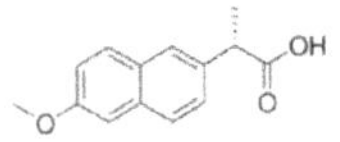	150	25	0.32
Nonylphenol	220.3	5.76	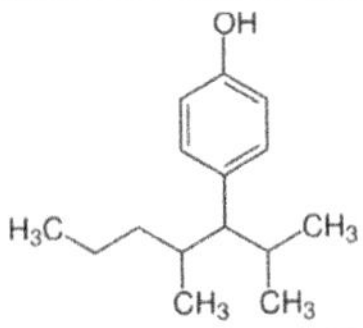	7	31.1	0.28
Omeprazole	345.4	2.23	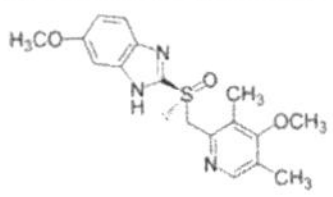	359	12.5	0.23
Oxazepam	286.7	2.24	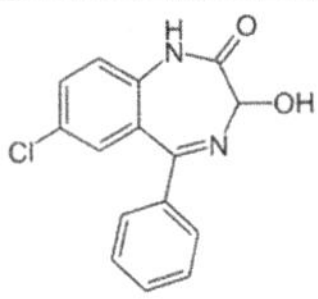	179	12.8	0.21
Ofloxacin	361.4	–0.39	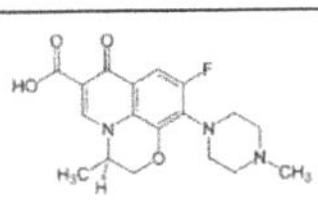	10,800	11.2	0.31
Progesterone	272.4	4.01	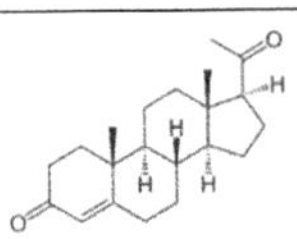	3.60	25.6	0.29
Paracetamol	151.1	0.46	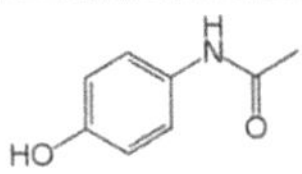	14,000	2.8	0.26

Primidone	218.2	0.91	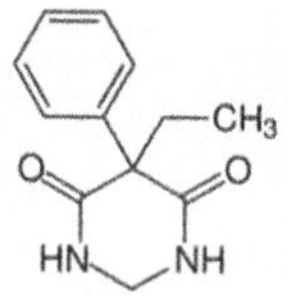	500	3.5	0.29
Propanolol	259.3	3.48	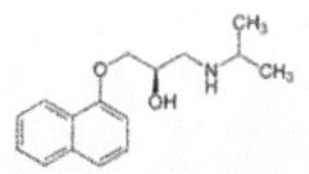	61.7	41	0.19
Sotalol	272.3	0.24	–	782	8	0.22
Sulfamethoxazole	253.2	0.89	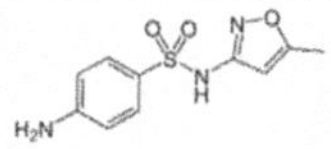	610	3.5	0.24
Tramadol	299.8	3.01	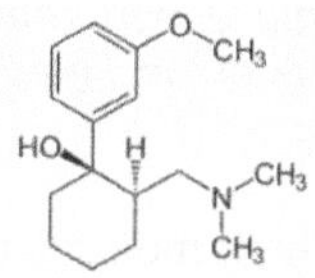	1,151	13	0.22
Triclocarban	315.6	4.34	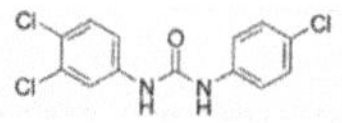	0.11	23.3	0.31
Triclosan	289.5	4.76	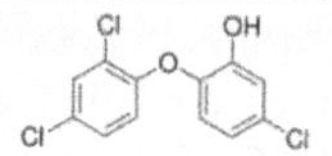	10	17.2	0.38
Trimethoprim	290.3	0.91	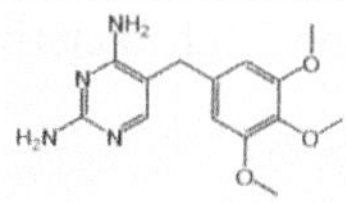	400	35	0.23
Valsartan	485.5	1.49	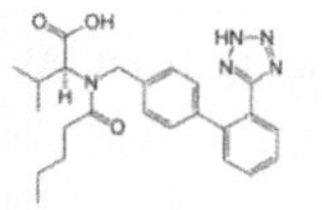	8,210	7	0.19
Venlafaxine	277.4	3.20	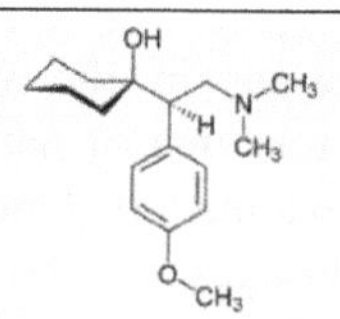	267	12.8	0.21

Table 2.10. *Freundlich parameters (K and 1/n) for some drugs*

Adsorption results obtained with Actiflo® Carb on PACs (Aquasorb MP23, Cabot-Norit SAE Super and Pulsorb RD 30) for micropollutant concentrations ranging from 500 to 1,000 ng.L^{-1} confirm the predictions.

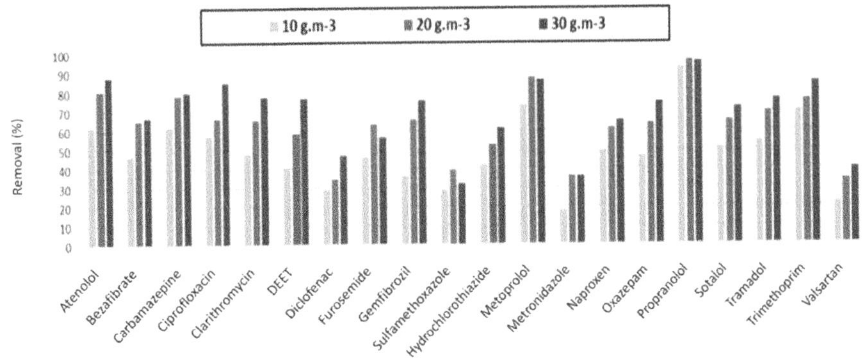

Figure 2.20. *Micropollutant removal performance with different dosages of Pulsorb RD 90 PAC. For a color version of this figure, see www.iste.co.uk/gaid/reuse2.zip*

Iohexol and iomeprol have high molecular weights, but because of their low log Kow and size, they are poorly adsorbed.

Gabapentin, iohexol and metformin are poorly adsorbed (<20%). Atenolol, tramadol, propranolol, carbamazepine, metoprolol and fenofibrate are well adsorbed (>70%), and among the best of all the other micropollutants. Their high log Kow gives them a high affinity for PAC. Triclosan, which has a symmetrical, coplanar and relatively small structure and is characterized by a high log Kow, is also well adsorbed.

PAC reactors can achieve removal efficiencies for a large number of micropollutants, with values ranging from 60 to 95%, with activated carbon doses varying from 10 to 20 mg.L^{-1}.

The performance of the PAC reactor is assessed on the basis of all the pesticides and other micropollutants present in the water to be treated. This is what differentiates this application from that of natural organic matter, where only one parameter (DOC), or even two parameters (color and DOC), had to be considered. It should be noted that PAC dosages can be increased depending on the DOC concentration in the water, which is an important factor influencing the removal of micropollutants through competitive adsorption.

2.4.3.3.3. Activated carbon/ultrafiltration membrane combination

An alternative to conventional reactors is to use PAC in a fluidized bed or to combine PAC with an ultrafiltration membrane.

The PAC/membrane combination, based on the Opaline® C process (Figure 2.21), is appealing when the PAC concentration to be injected does not require high doses, and when an initial removal of microorganisms is also required. The PAC contact tank is installed in the same way as conventional reactors, with an equivalent time.

Membrane pore size (cut-off) is determined by the size of the elements to be extracted from the water. Ultrafiltration is the most frequently used process for water clarification and disinfection. Ultrafiltration removes suspended solids, turbidity and various microorganisms (bacteria, viruses, parasites, etc.).

Operation of the PAC/UF plant involves two successive stages: adsorption on PAC in a special tank and membrane filtration. On leaving the contact tank, the mixture is directed to the ultrafiltration unit. Part of the mixture (5–20%) containing PAC is recycled to the top of the contact tank. This mode of operation creates turbulent movement on the membrane surface, thus reducing clogging.

Recirculation also concentrates and increases the residence time of the PAC in the system. The other part of the flow is filtered by the membrane in an internal-external direction, operating in tangential mode. The ultra-filtered water, known as permeate, is then directed to the filtered water tank.

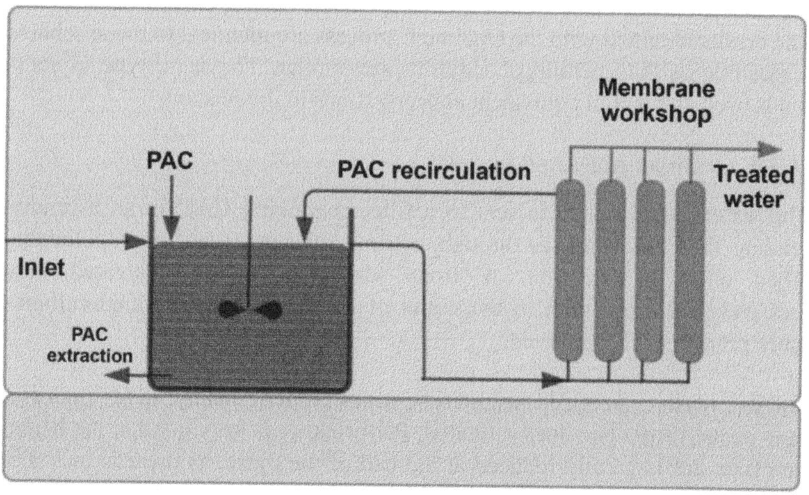

Figure 2.21. *Opaline® C process: combination of powdered activated carbon and pressurized ultrafiltration membrane*

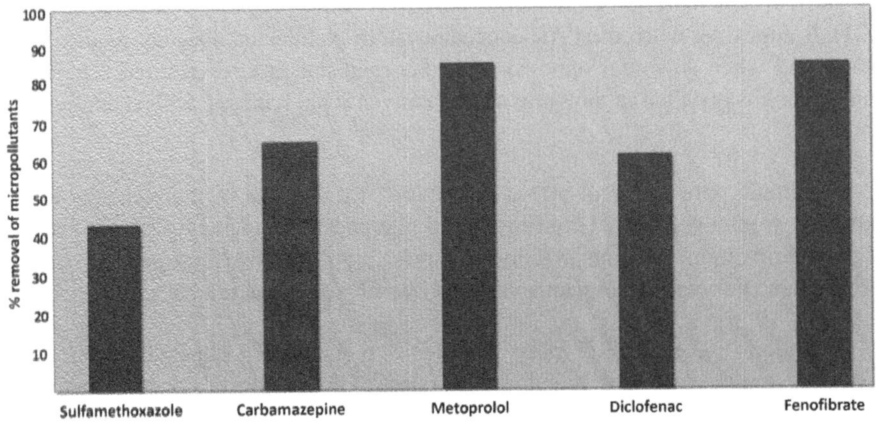

Figure 2.22. *Removal of micropollutants with Opaline® C process (Veolia)*

Activated carbon removes micropollutants as well as DOC, while the ultrafiltration membrane has no effect on micropollutant removal. Figure 2.22 shows an example of micropollutant removal with the Opaline® C process.

The results obtained with the Opaline C process are identical to those achieved in PAC reactors, as the operating conditions are similar. The same type of activated carbon is used and it is in equivalent concentrations in the reactor.

2.4.3.3.4. Granular activated carbon

The water to be treated is sent to a filter containing GAC in an ascending or descending flow. As it passes through, organic substances and micropollutants are adsorbed onto the material. A front adsorption layer is created, moving progressively from the inlet to the outlet of the filter. The activated carbon thus becomes progressively saturated.

For this reason, the GAC needs to be replaced periodically. GAC performance changes as the media becomes saturated. Performance is very good at the beginning of the cycle, but is greatly reduced at the end of the cycle, as there is only a small volume of unsaturated GAC still available in relation to the load capacity. For this reason, GAC filters require regular monitoring, and the media capacity must be controlled to avoid micropollutant leaks.

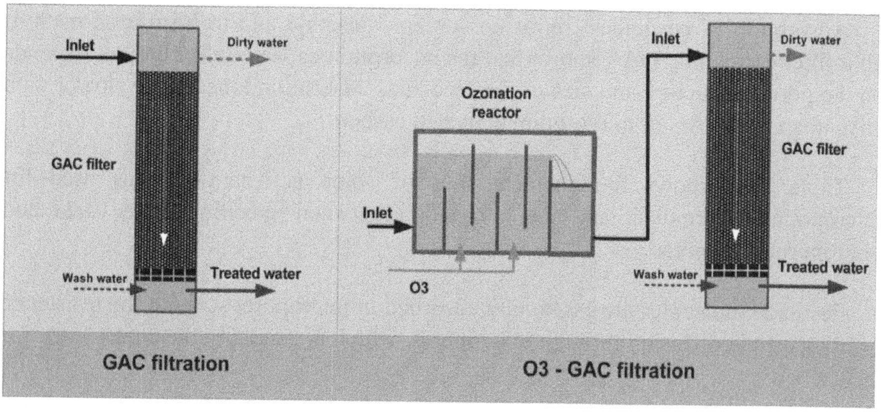

Figure 2.23. *Removal of micropollutants using GAC or in combination with ozone*

2.4.3.3.5. Adsorption of pesticides on a GAC filter

Parameters	**Granular activated carbon**			
Supplier	**Jacobi**	**Jacobi**	**Cabot-Norit**	**Chemviron**
Name	Aquasorb NC 60	Aquasorb 2000	Norit Row 0.8	12 × 40
Origin	Coconut	Coal	Peat	Coal
Grain diameter (mm)	1.0	1.0	0.9	1.1
Specific surface area ($m^2.g^{-1}$)	1,200	1,000	915	1,100
Total pore volume ($cm^3.g^{-1}$)	0.590	–	0.648	0.646
Micropore volume ($cm^3.g^{-1}$)	0.513	–	0.388	0.466
Macroporous volume ($cm^3.g^{-1}$)	0.242	–	0.375	0.646
Microporosity %	87	–	59.9	72.2
Average pore diameter (nm)	0.59	–	0.55	0.58
Density ($kg.m^{-3}$)	0.783	–	0.646	0.579
Bulk density ($kg.m^{-3}$)	0.97	0.95	0.85	0.92

Table 2.11. *Characteristics of some granular activated carbons used for micropollutant removal*

Adsorption of pesticides (atrazine, atrazine-desethyl and triflusulfuron-methyl) on a filter containing GAC shows that the adsorption capacity of pesticides depends on the porous structure and size of the media, as adsorption kinetics are slower with GAC than with PAC or micro-grain activated carbon.

Table 2.11 shows the characteristics of some activated carbons used for micropollutant removal, and the correlation between specific surface area and microporosity is noted.

Pesticide molecules are essentially adsorbed in micropores without the resistance of intraparticle diffusion through macropores, which is generally the control step for GACs.

2.4.3.3.6. Adsorption of drug residues and other endocrine disruptors on a GAC filter

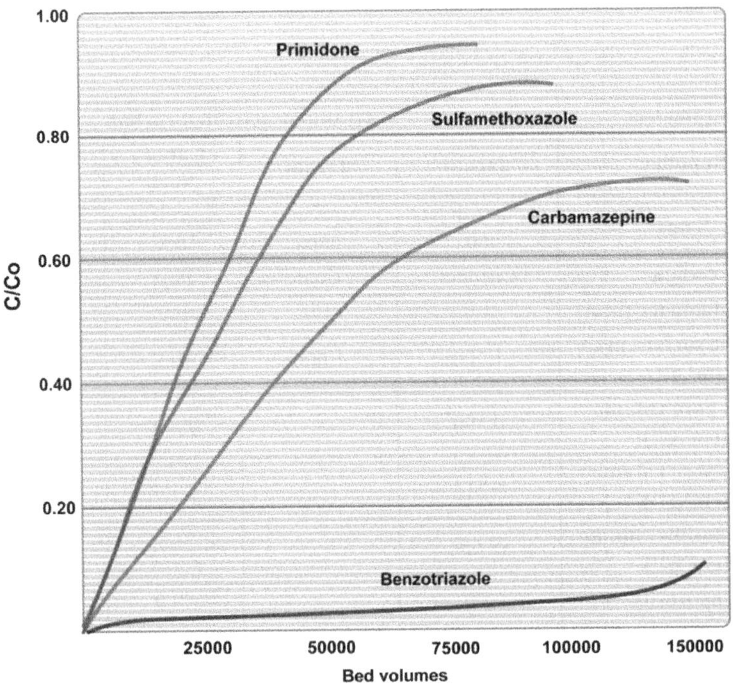

Figure 2.24. *Adsorption of four drugs on a GAC filter (Aquasorb 2000)*

The water fed to GAC filters is usually water that has been coagulated, flocculated, decanted and filtered. A residual amount of dissolved organic substances (DOC) remains in addition to the micropollutants that were not eliminated in the previous stages, leading to competitive adsorption that can reduce the activated carbon's capacity to remove micropollutants. The compounds present in natural organic matter are low-molecular-weight acid and/or neutral, and are competitors for micropollutants.

Figure 2.24 shows adsorption results on a GAC filter for three drug residues (carbamazepine, primidone and sulfamethoxazole) and one industrial residue (benzotriazole). The GAC filter is made of Aquasorb 2000 activated carbon, with a contact time of 10 min, operating at a speed of 8 m.h^{-1}. Micropollutant inlet concentrations range from 5 to 10 µg.L^{-1}. Benzotriazole appears to be the best adsorbed, while GAC saturates primidone after 75,000 bed volumes (BV). Sulfamethoxazole behaves almost identically to primidone. Carbamazepine saturates the GAC by 50% after 50,000 BV.

However, the capacity of carbon to adsorb micropollutants in water is a function of the initial concentration of micropollutants: the lower the initial concentration in natural water, the lower the observed adsorption capacity.

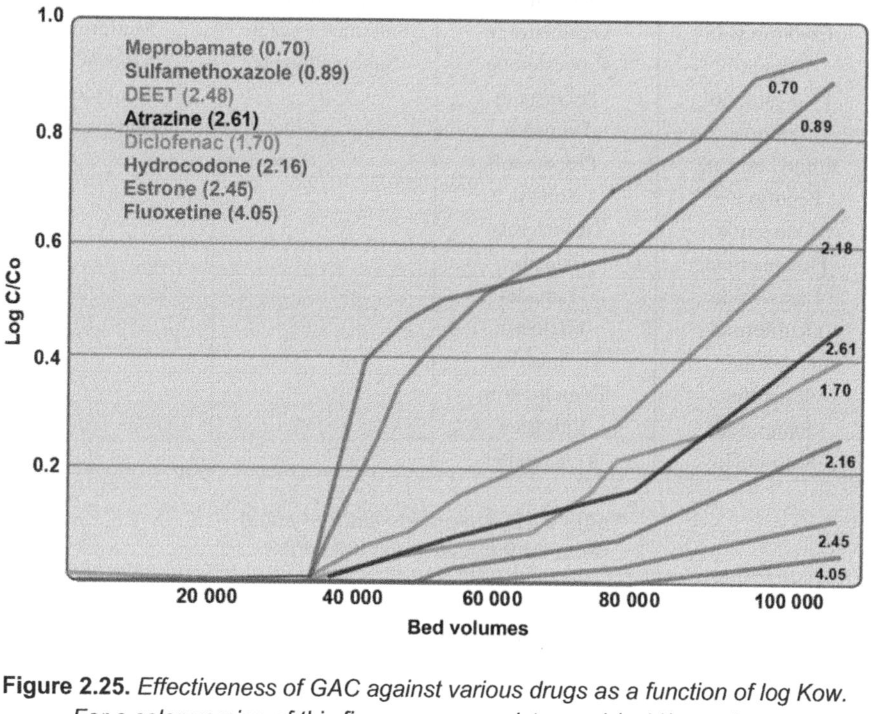

Figure 2.25. *Effectiveness of GAC against various drugs as a function of log Kow. For a color version of this figure, see www.iste.co.uk/gaid/reuse2.zip*

Figure 2.25 shows the effectiveness of Cabot–Norit activated carbon (12 × 40) against various drugs characterized by different log Kow values. This figure confirms that a molecule with a high log Kow would be better adsorbed than another molecule with a low to negative log Kow. This is a trend, but not a general rule.

Table 2.12 shows the adsorption efficiency of activated carbon in relation to selected drug residues.

Residues of well-adsorbed drugs		**Residues of medium-adsorbed drugs**	**Residues of poorly adsorbed drugs**
Amlodipine	Levonorgestrel	Bisphenol A	Caffeine
Atenolol	Losartan	Ciprofloxacin	Fosfomycin
Bezafibrate	Metoprolol	Cyclophosphamide	Hydroxycarbamide
Carbamazepine	Naxoprene	Doxycycline	Hydrochlorothiazide
Clarithromycin	Nonylphenol	Heptaminol	Iopamidol
Chlorambucil	Omeprazole	Fosfamide	Iobitridol
Diclofenac	Oxazepam	Meprobamate	Iohexol
DEET	Octylphenol	Paracetamol	Levodopa
Erythromycin	Oxylbenzone	Sulfamethoxazole	Metformin
Estradiol	Progesterone	Sulfamethazine	Ofloxacin
17-α-estradiol	Pravastatin		Ranitidine
Estriol	Primidon		Gabapentin
Ethinyl estrone	Propanolol		
Fenofibrate	Solalol		
Fluoxetine	Testosterone		
Fluvoxamine	Tilosine		
Furosemide	Tramadol		
Gemfibrozil	Triclosan		
Galaxolide	Triclocarban		
Ibuprofen	Trimethoprim		
Iopromide	Valsartan		
Ketoprofen	Venlafaxine		

Table 2.12. *Adsorption state of some drug residues on activated carbon*

2.4.4. *Ozone/activated carbon combination*

Ozone is known to be an effective solution for oxidizing micropollutants, but has the disadvantage of forming by-products, as Veolia has shown, which could potentially be EDs in the same way as the parent substance.

2.4.4.1. *Ozone/PAC*

An O_3/PAC combination is possible by blowing ozone into the reactor itself, thus avoiding the formation of by-products while ensuring a significant increase in removal efficiency. The principle is that the O_3 acts on the adsorbent material and not on the micropollutant to be removed. This process achieves removal rates of over 90% for the vast majority of micropollutants. The introduction of ozone in direct contact with the activated carbon leads to a modification of the surface of the activated carbon, creating surface functions that increase the adsorption capacity of the adsorbent medium. Micropollutants will thus bind to the adsorption sites on the activated carbon and to the surface functions. The result is better removal of micropollutants at the same or even lower dose of activated carbon. It is thus possible to significantly reduce PAC doses by using ozone, while maintaining identical process performance. No organic by-products or bromates are formed.

The results obtained with an O_3 dosage of 1–2 mg.L^{-1}, a PAC contact time of 10 min and PAC dosages of 5–10 mg.L^{-1} showed a gain of around 10–15% in the average removal performance of the various micropollutants, with average removal rates of around 85–92%, instead of the 70–80% observed with PAC.

Substances	O_3/PAC (Aquasorb MP23)			O_3/PAC (Norit SAE Super)		
	Inlet	Outlet	% elim.	Inlet	Outlet	% elim.
Diclofenac, µg.L^{-1}	0.839	0.149	82	1.093	0.034	97
Carbamazepine, µg.L^{-1}	0.982	0.024	98	1.288	0.014	99
Hydrochlorothiazide, µg.L^{-1}	2.333	0.077	97	3.658	0.146	95
Metoprolol, µg.L^{-1}	0.033	0.010	69	0.050	0.010	76
Sulfamethoxazole, µg.L^{-1}	0.195	0.045	77	0.256	0.030	88
Atenolol, µg.L^{-1}	0.150	0.013	90	0.142	0.018	82
Ethylestradiol, µg.L^{-1}	5.211	0.108	98	11.075	0.301	97

Table 2.13. *Performance of two PACs combined with ozone*

It can be seen that micropollutant removal performance is fairly similar for these drug residues.

Optimum removal of micropollutants is thus achieved using this O_3/PAC reactor combination, without releasing any by-products into the receiving environment. Results obtained at a second site confirm the effectiveness of this combination, applied under the same operating conditions described above. Efficiencies of over 80% were obtained for all the molecules analyzed, with values in excess of 98% for carbamazepine, diclofenac, hydrochlorothiazide and sulfamethoxazole.

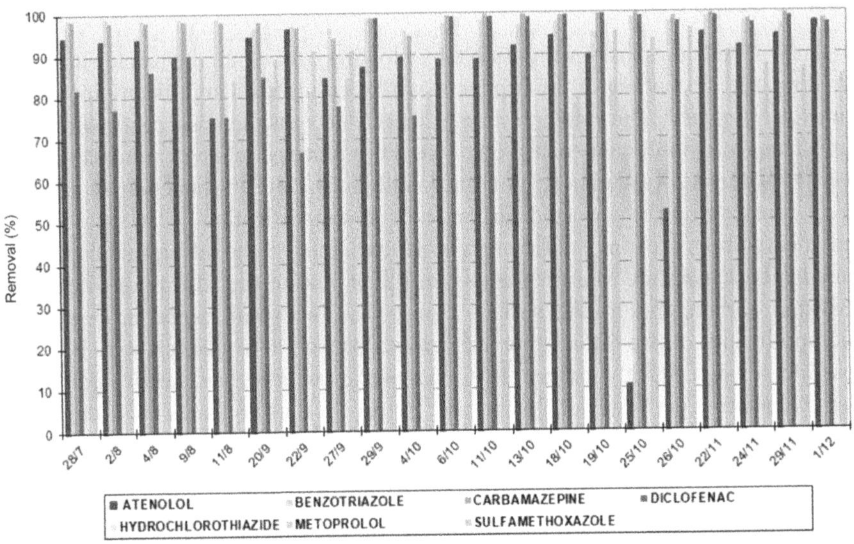

Figure 2.26. *Summary of Actiflo® Carb performance with O_3/PAC reactor coupling. For a color version of this figure, see www.iste.co.uk/gaid/reuse2.zip*

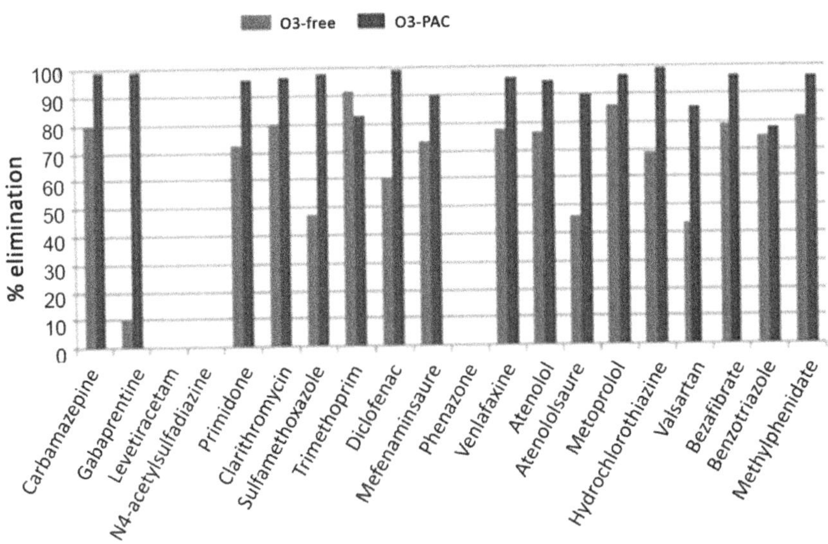

Figure 2.27. *Removal of micropollutants with the in situ O_3 /PAC combination (O_3: 2 g.m^{-3}; PAC (Aquasorb MP23): 10 g.m^{-3}). For a color version of this figure, see www.iste.co.uk/gaid/reuse2.zip*

Optimum removal of micropollutants (Figure 2.27) is thus achieved with this in situ O_3 and PAC reactor combination, without releasing any by-products into the receiving environment. Compounds with low or moderate removal rates by PAC alone (due to the low activated carbon doses), such as gabapentin (10%), atenolol (40%) and valsartan (40%), see their removal efficiency rise to over 80% in the presence of in situ O_3.

This efficiency can be significantly improved by adjusting the dosages of the two reagents, while controlling the ozone dosage to avoid the formation of by-products.

– Mass removal for all molecules analyzed rose to over 91% (versus 75% without ozone).

– Average removal for all molecules analyzed rose to 89% (versus 77% without ozone).

Although the applied ratios of 0.5–0.6 mg O_3.mg^{-1} DOC are in line with those previously described for oxidizing micropollutants with O_3, the mechanisms of the O_3/PAC combination are different, as the O_3 is no longer injected for ozonation purposes, but rather to modify the surface of the activated carbon.

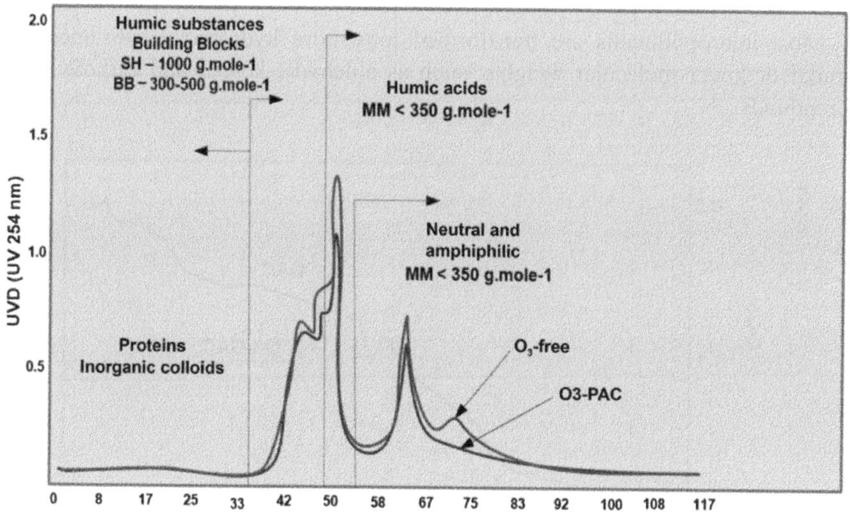

Figure 2.28. *LC UVD of samples treated with Actiflo® Carb with and without activated carbon ozonation. For a color version of this figure, see www.iste.co.uk/gaid/reuse2.zip*

As ozone is injected directly into the PAC contact tank, which has a concentration of between 1.5 and 2 g.L^{-1}, there is a direct reaction between the oxidant (O_3) and the reducer (PAC) to the detriment of organic matter and micropollutants. This is why the ozonation reaction does not affect micropollutants, as O_3/PAC reaction kinetics are much faster. Furthermore, residual ozone at a concentration of 0 mg/L has always been found to be zero in pore water. To complement the micropollutant analyses, we compared the SEC-UVD chromatograms for the two scenarios, with and without ozone injected into the carbon reactor. Analyses of samples taken at the inlet/outlet of the Actiflo® Carb show (Figure 2.28) that humic substances are indeed removed by the addition of activated carbon. The injection of O_3 into the PAC reaction tank results in a systematic reduction in humic substance peaks, compared with those obtained with Actiflo® Carb (without O_3). The sample thus treated contains a lower concentration of organic carbon. This can be explained by the improved adsorption efficiency of PAC oxidized in situ by ozone, without however generating new peaks that would have been assimilated to oxidation by-products.

2.4.4.2. *O_3/granular activated carbon*

Ozone doses of 0.8–1.2 g O_3.g^{-1} DOC and a hydraulic retention time (HRT) of 12–20 min are sufficient to remove a wide range of micropollutants.

Most micropollutants are transformed into more hydrophilic substances with similar or lower molecular weights, such as aldehyde, ketone and carboxylic acid compounds.

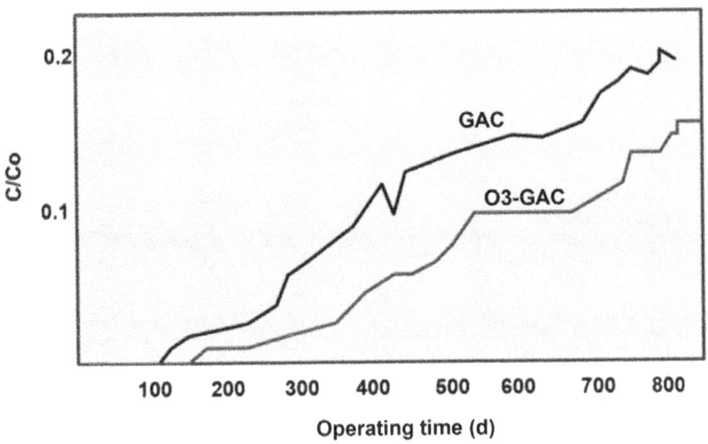

Figure 2.29. *Evolution of atrazine elimination, with ozonation alone and with O_3/GAC*

The aim of an ozone/GAC filter combination is to initiate and promote the chemical decomposition of organic substances by ozone, then to ensure their biological degradation on the GAC filter.

Figure 2.29 shows the evolution of the service life of a GAC filter fed with surface water containing organic carbon (DOC: 2.4–3.0 mg.L^{-1}) and a pesticide (atrazine: 0.5–0.7 µg.L^{-1}). Hydraulic contact time in the GAC filter is 10 min.

Figure 2.29 also shows the evolution of the service life of a GAC filter fed with the same water, but treated with an O_3/GAC combination. The contact time in the O_3 tanks is 6 min. Hydraulic contact time in the GAC filter is 10 min.

Atrazine leakage in excess of 0.1 µg.L^{-1} is achieved with GAC after 400 days of operation. It is highly likely that the organic matter present blocked some of the adsorption sites, preventing atrazine from being retained any longer.

With the O_3/GAC combination, atrazine leakage is only observed after 700 days of operation. The oxidation of DOC by O_3 and its biodegradation on GAC certainly freed up adsorption pores, enabling greater retention of atrazine. Furthermore, the by-products of atrazine oxidation, desethylatrazine and DIA, are more biodegradable and less adsorbable than atrazine; it is therefore rather difficult to predict the leakage time of these two metabolites. Moreover, as a function of water temperature, the biodegradation function is more effective in warmer periods than in colder ones.

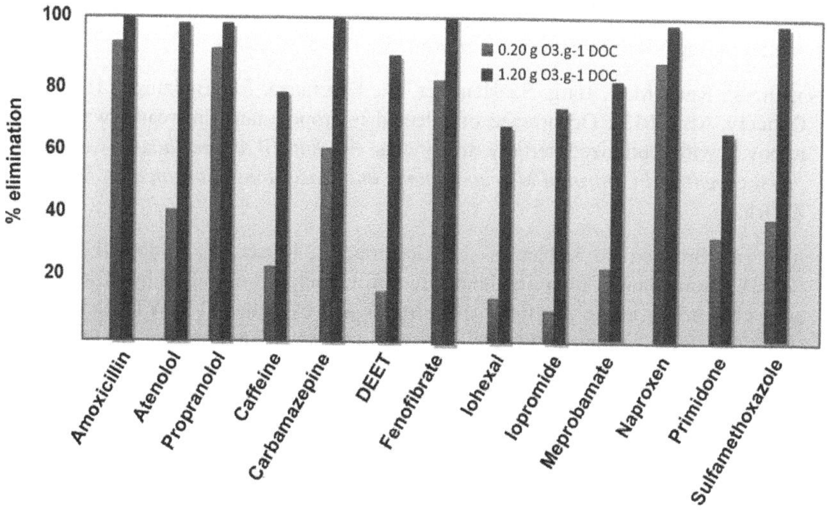

Figure 2.30. *Removal of some drugs with O_3/GAC according to two O_3/DOC ratios. For a color version of this figure, see www.iste.co.uk/gaid/reuse2.zip*

Results obtained on an O_3/GAC combination with a GAC operating essentially in biological mode (Figure 2.30) show that the formation of biodegradable by-products requires an O_3/ DOC ratio greater than 1 for a large number of compounds.

Biological filtration on GAC, based on a contact time of 15 min, achieves 80% efficiency on both parent molecules and by-products. Biological filtration on activated carbon removes more than 80% of drug residues, provided their concentration is at least 10 times their limit of quantification (LOQ) prior to GAC filtration.

Gabapentin, on the other hand, is only 53% removed, due to its highly hydrophilic nature (log Kow: –1.37). Furthermore, biological filtration on GAC removes between 25 and 35% of DOC at temperatures above 15°C.

2.5. References

Abellan, M.N., Gebhardt, W., Schroder, H.F. (2008). Detection and identification of degradation products of sulfamethoxazole by means of LC/MS and -MS(n) after ozone treatment. *Water Sci. Technol.*, 58(9), 1803–1812.

Antoniou, M.G., Hey, G., Vega, S.R., Spiliotopoulou, A., Fick, J., Tysklind, M., Jansen, J., Andersen, H.R. (2013). Required ozone doses for removing pharmaceuticals from wastewater effluents. In *8th IWA Specialized Conference on "Assessment and Control of Micropollutants and Hazardous Substances in Water"*. IWA, Zurich.

Bellona, C.J., Drewes, J.E., XU, P., Amy, G. (2004). Factors affecting the rejection of organic solutes during NF/RO treatment. *Wat. Res.*, 38(12), 2795–2809.

Besnault, S., Ruel, M.S., Baig, S., Heiniger, B., Esperanza, M., Budzinski, H., Miège, C., Coquery, M. (2013). Occurrence of selected micropollutants in treated wastewater and removal with optimized tertiary treatments. In *8th IWA Specialized Conference on "Assessment and Control of Micropollutants and Hazardous Substances in Water"*. IWA, Zurich.

Boehler, M., Brocker, S., Kobler, S., Le Goaziou, Y., Moser, R., Rettby, R., Rigault, M. (2012). Elimination of micro pollutants out of municipal wastewater by means of powder activated carbon in the Actiflo® Carb process and by ozone at WWTP Schönau, Cham (GVRZ). Final Report, Federal Office for the Environment (FOEN), Alpha Umwelttechnik, Veolia, GVRZ, Labosafe, Hunziker-Betatech, Eawag.

Broseus, R., Vincent, S., Aboulfadl, K., Daneshvar, A., Sauvé, S., Barbeau, B., Prevost, M. (2009). Ozone oxidation of pharmaceuticals, endocrine disruptors and pesticides during water treatment. *Water Research*, 43, 4707–4717.

Comerton, M.A., Andrews, C.R., Bagley, M.D., Hao, C. (2008). The rejection of endocrine disrupting and pharmaceutically active compounds by NF and RO membranes as a function of compound and water matrix properties. *J. Mem. Sci.*, 313, 323–335.

Corwin, C.J. and Summer, R.S. (2012). Controlling trace organic contaminants with GAC adsorption. *Jour. Amer. Water Works Assoc.*, 104, 36–47.

Cotton, G., Feliers, C., Kecili, K., Colon, A.F. (2019). Performance du traitement membranaire (NDF et OIBP) : rétention des micropolluants et enjeux énergétiques. *Techniques sciences méthodes*, 4(104), 29–35.

Cougnaud, A., Faur, C., Le Cloirec, P. (2005). removal of pesticides from aqueous solution: Quantitative relationship between activated carbon characteristics and adsorption properties. *Environmental Technology*, 26, 852–866.

Dantas, R., Canterino, M., Marotta, R., Sansa, C., Esplugas, S., Andreozzi, R. (2007). Bezafibrate removal by means of ozonation: Primary intermediates, kinetics and toxicity assessment. *Water Research*, 41(12), 2525–2532.

Dolar, D., Vukovic, A., Asperger, D., Kosutic, K. (2012). Efficiency of RO/NF membranes at the removal of veterinary antibiotics. *Water Science & Technology*, 65(2), 317–323.

Drault-Pezard, C. (2012). Ozonation et procédés d'oxydation avancée : solutions d'avenir pour traiter les polluants émergents. *L'eau, l'industrie, les nuisances*, 351, 55–60.

Drewes, J., Heberer, T., Redderson, K. (2002). Fate of pharmaceuticals during indirect potable reuse. *Water Sci. Technol.*, 46(3), 73–80.

El Najjar, N.H., Touffet, A., Deborde, M., Journel, R., Vel Leitner, N.K. (2013). Levofloxacin oxidation by ozone and hydroxyl radicals: Kinetic study, transformation products and toxicity. *Chemosphere*, 93(4), 604–611.

Gaid, K. and Sauvignet, P. (2012). Elimination des micropolluants en eaux usées avec réacteur charbon actif en poudre, (Actiflo® Carb) et suivi de la qualité du traitement. In *Journées Informations Eaux*, Poitiers.

Gaid, K. and Sauvignet, P. (2013). High-rate PAC reactor combined with ozone oxidation for enhanced micro-pollutants removal from urban waste water. In *8th IWA Specialized Conference on "Assessment and Control of Micropollutants and Hazardous Substances in Water"*. IWA, Zurich.

Gaid, K. and Sauvignet, P. (2015). Substances organiques, perturbateurs endocriniens, micropolluants : une solution simple le Filtraflo carb. *L'eau, l'industrie, les nuisances*, 382, 89–93.

Gaid, K., Sauvignet, P., Marlin, P. (2014). Use of PAC (powder activated carbon) combined with Actiflo & UF membrane for enhanced NOM removal from a high DOC (dissolved organic carbon) water source. In *Organic Water Removal in Drinking Water Conference*, Nottingham University, Nottingham.

Gaid, K., Sauvignet, P., Bourdon, F., Ingrand, V., Rigot, M. (2015). Les réacteurs CAP "Multiflo-Actiflo" : la solution pour l'éliminations des perturbateurs endocriniens et des substances dangereuses. *L'eau, l'industrie, les nuisances*, 353, 68–75.

von Gunten, U. (2003). Ozonation of drinking water: Part I. Oxidation kinetics and product formation. *Water Research*, 37, 1443–1467.

Gur-Reznik, S., Koren-Menashe, I., Heller-Grossman, L., Rufel, O., Dosoretz, C.G. (2011). Influence of seasonal and operating conditions pf the rejection of pharmaceutical active compounds by RO and NF membranes. *Desalination*, 277, 250–256.

Guyard, C. (2012). Dégradation de la ressource en eau : faire face aux polluants émergents. *L'eau, l'industrie, les nuisances*, 356, 45–50.

Hollender, G. (2009). Elimination of organic micropollutants in a municipal wastewater treatment plant upgraded with a full-scale post ozonation followed by sand filtration. *Environmental Science and Technology*, 43(20), 7862–7869.

Huber, M.M., Canonica, S., Park, G.Y., von Gunten, U. (2003). Oxidation of pharmaceuticals during ozonation and advanced oxidation processes. *Environmental Sciences & Technology*, 37(5), 1016–1024.

Kimura, K., Amy, G., Drewes, J.E., Watanabe, Y. (2003). Adsorption of hydrophobic compounds onto NF/RO membranes: An artifact on over estimation of rejection. *J. Mem. Sci.*, 221, 89–101.

Kisoa, Y., Nishimuraa, Y., Kitaob, T., Nishimurab, K. (2000). Rejection properties of non-phenylic pesticides with nanofiltration membranes. *Journal of Membrane Science*, 171, 229–237.

Košutić, K. and Kunst, B. (2002). Removal of organics from aqueous solutions by commercial RO and NF membranes of characterized porosities. *Desalination*, 142, 47–56.

Košutić, K., Dolar, D., Ašperger, D., Kunst, B. (2007). Removal of antibiotics from a model wastewater by RO/NF membranes. *Sep. Purif. Technol.*, 53, 244–249.

Larcher, S., Delbès, G., Robaire, B., Yargeau, V. (2012). Degradation of 17α-ethinylestradiol by ozonation – Identification of the by-products and assessment of their estrogenicity and toxicity. *Environment International*, 39, 66–72.

Löwenberg, J., Zenker, A., Baggenstos, M., Koch, G., Kazner, C., Wintgens, T. (2013). Comparison of two PAC/UF processes for the removal of micropollutants from wastewater treatment plant effluent: Process performance and removal efficiency. In *8th IWA Specialized Conference on "Assessment and Control of Micropollutants and Hazardous Substances in Water"*. IWA, Zurich.

Luster-Teasley, S.L., Yao, J.J., Herner, H.H., Trosko, J.E., Masten, S.J. (2002). Ozonation of chrysene: Evaluation of byproduct mixtures and identification of toxic constituent. *Environmental Science and Technology*, 36, 869–876.

Margot, J., Magnet, A., Thonney, D., Chèvre, N., de Alencastro, F., Rossi, L. (2011). Traitement des micropolluants dans les eaux usées. Rapport final sur les essais pilotes à la STEP de Vidy. Ville de Lausanne, Lausanne.

Margot. J., Kienle, C., Magnet, A., Weil, M., Rossi, L., Felipe de Alencastro, L., Abegglen, C., Thonney, D., Chèvre, N., Schärer, M. et al. (2013). Treatment of micropollutants in municipal wastewater: Ozone or powdered activated carbon? *Science of the Total Environment*, 461(462), 480–498.

McDowell, D.C., Huber, M.M., Wagner, M., von Gunten, U., Ternes, T.A. (2005). Ozonation of carbamazepine in drinking water: Identification and kinetic study of major oxidation products. *Environmental Science & Technology*, 39(20), 8014–8022.

Mestankova, H., Bramaz, N., Schirmer, K., Canonica, S., Linden, K., von Gunten, U. (2013). Advanced oxidation processes: Toxicity of transformation products. In *8th IWA Specialized Conference on "Assessment and Control of Micropollutants and Hazardous Substances in Water"*. IWA, Zurich.

Nghiem, L.D. (2005). Removal of emerging trace of organic contaminants by naofiltration and reverse osmosis. PhD Thesis, University of Wallongong, Wallongong.

Nghiem, L.D., Schäfer, A.I., Elimelech, M. (2005). Pharmaceutical retention mechanisms by nanofiltration membranes. *Environ. Sci. Technol.*, 39, 7698–7705.

Rattier, M., Reungoat, J., Joss, A., Gernjak, W., Keller, J. (2013). Evaluation of the role of adsorption and biodegradation of micropollutants during biological activated carbon filtration. In *8th IWA Specialized Conference on "Assessment and Control of Micropollutants and Hazardous Substances in Water"*. IWA, Zurich.

Rosal, R., Rodríguez, A., Perdigón-Melón, J.A., Petre, A., García-Calvo, E., Gómez, M.J. et al. (2010). Occurrence of emerging pollutants in urban wastewater and their removal through biological treatment followed by ozonation. *Water Research*, 44, 578–588.

Ruhl, A.S., Altmann, J., Meinel, F., Zietzschmann, F., Sperlich, A., Jekel, M. (2013). Enhancing the efficiency of powdered activated carbon for micro pollutant removal by immobilization in a fixed bed. In *8th IWA Specialized Conference on "Assessment and Control of Micropollutants and Hazardous Substances in Water"*. IWA, Zurich.

Schaar, P. (2010). Micropollutant removal during biological wastewater treatment and a subsequent ozonation step. *Environmental Pollution*, 158(5), 1399–1404.

Sein, M.M., Zedda, M., Tuerk, J., Schmidt, T.C., Golloch, A., Von Sonntag, C. (2008). Oxidation of diclofenac with ozone in aqueous solution. *Environ. Sci. Technol.*, 2, 6656–6662.

Seung-Woo, N., Dae-Jin, Ch., Seung-Kyu, K., Namguk, H., Kyung-Duk, Z. (2014). Adsorption characteristics of selective hydrophilic and hydrophobic micropollutants in water using activated carbon. *Journal of Hazardous Materials*, 270, 144–152.

Snyder, S.A., Adham, S., Redding, A.M., Cannon, F., DeCarolis, J., Oppenheimer, J., Wert, E.C., Yoon, Y. (2007a). Role of membranes and activated carbon in the removal of endocrine disruptors and pharmaceuticals. *Desalination*, 202(1/3), 156–181.

Snyder, S.A., Wert, E.C., Lei, H., Westernhoff, P., Yoon, Y. (2007b). Removal of EDCs and pharmaceuticals in drinking and reuse treatment processes. Report, A.W.W.A. Research Foundation, Denver.

Snyder, S.A., Adham, S., Redding, A.M. (2007c). Role of membranes and activated carbon in the removal of endocrine disruptors and pharmaceuticals. *Desalination*, 202(1/3), 156–181.

Ternes, H. (2003). Ozonation: A tool for removal of pharmaceuticals, contrast media and musk fragrances from wastewater. *Water Research*, 37(8), 1976–1982.

Tootchi, L., Seth, R., Tabe, S., Yang, P. (2013). Transformation products of pharmaceutically active compounds during drinking water ozonation. *Water Supply*, 13(6), 1576–1582.

Van der Bruggen, B. and Vandecasteele, C. (2002). Modelling of the retention of uncharged molecules with nanofiltration. *Water Research*, 36, 1360–1368.

Van der Bruggen, B., Schaep, J., Wilms, D., Vandecasteele, C. (1999). Influence of molecular size, polarity and charge on the retention of organic molecules by nanofiltration. *Journal of Membrane Science*, 156, 29–41.

Verliefde, A.R.D., Heijman, S.G.J., Cornelissen, E.R., Amy, G., Van der Bruggen, B., van Dijk, J.C. (2007). Influence of electrostatic interactions on the rejection with NF and assessment of the removal efficiency during NF/GAC treatment of pharmaceutically active compounds in surface water. *Water Research*, 41(15), 3227–3240.

Wert, E.C., Rosario-Ortiz, F.L., Drury, D.D., Snyder, S.A. (2007). Formation of oxidation byproducts from ozonation of wastewater. *Water Research*, 41, 1481–1490.

Wert, E.C., Rosario-Ortiz, F.L., Snyder, S.A. (2009). Effect of ozone exposure on the oxidation of trace organic contaminants in wastewater. *Water Research*, 43, 1005–1014.

Yangali-Quintanilla, V., Verliefde, A., Kim, T.-U., Sadmani, A., Kennedy, M., Amy, G. (2009). Artificial neural network models based on QSAR for predicting rejection of neutral organic compounds by polyamide nanofiltration and reverse osmosis membranes. *Journal of Membrane Science*, 342, 251–262.

Yargeau, V., Segura, P.A., Kaplan, P. (2013). Nature and estrogenic activity of ozonation transformation products of estrone. In *8th IWA Specialized Conference on "Assessment and Control of Micropollutants and Hazardous Substances in Water"*. IWA, Zurich.

Ying, G., Kookanai, R., Waite, T.D. (2003). Endocrine disrupting chemicals and pharmaceutical and personal care products in reclaimed water in Australia. Report, Australian Water Conservation and Reuse Research Program.

Yu, Z., Peldszus, S., Huck, P.M. (2009). Adsorption of selected pharmaceuticals and an endocrine disrupting compound by granular activated carbon. 1. Adsorption capacity and kinetics. *Environmental Science and Technology*, 43(5), 1467–1473.

Yu, J., Lv, L., Lan, P., Zhanga, S., Pana, B., Zhang, W. (2012). Effect of effluent organic matter on the adsorption of perfluorinated compounds onto activated carbon. *Journal of Hazardous Materials*, 225(226), 99–106.

Yoon, Y., Westerhoff, P., Snyder, A., Wert, E.C. (2006). Nanofiltration and ultrafiltration of endocrine disrupting compounds, pharmaceuticals and personal care products. *J. Membr. Sci.*, 270, 88–100.

3

Microfiltration and Ultrafiltration Membranes

Membrane technologies are considered an integral part of municipal wastewater treatment processes.

A membrane is a semi-permeable barrier that selectively separates compounds present in water under the effect of a pressure gradient. There are a large number of membranes on the market with different characteristics.

There are different types of membrane, classified according to the type and size of substances they can separate from water. Generally speaking, the smaller the size and molecular weight of the substances or ions to be separated, the higher the operating pressures of the membrane system, and in turn, the more efficient the upstream treatment. The best choice of membrane is therefore not the most selective one, but the one that is sufficiently selective for the objectives to be achieved, and offers optimum operating and investment costs.

Membrane technologies fall into two categories:

– low-pressure membranes: microfiltration (MF) and ultrafiltration (UF);

– high-pressure membranes: nanofiltration (NF) and reverse osmosis (low-pressure and high-pressure RO).

Depending on their category, they act as impenetrable barriers to suspended solids, colloids, bacteria, parasites, salts, organic substances and micropollutants.

A recurring problem with membranes is the clogging that occurs during the filtration process, followed by an increase in transmembrane pressure (TMP) to maintain a constant flux, or followed by a decrease in flux to maintain a constant TMP.

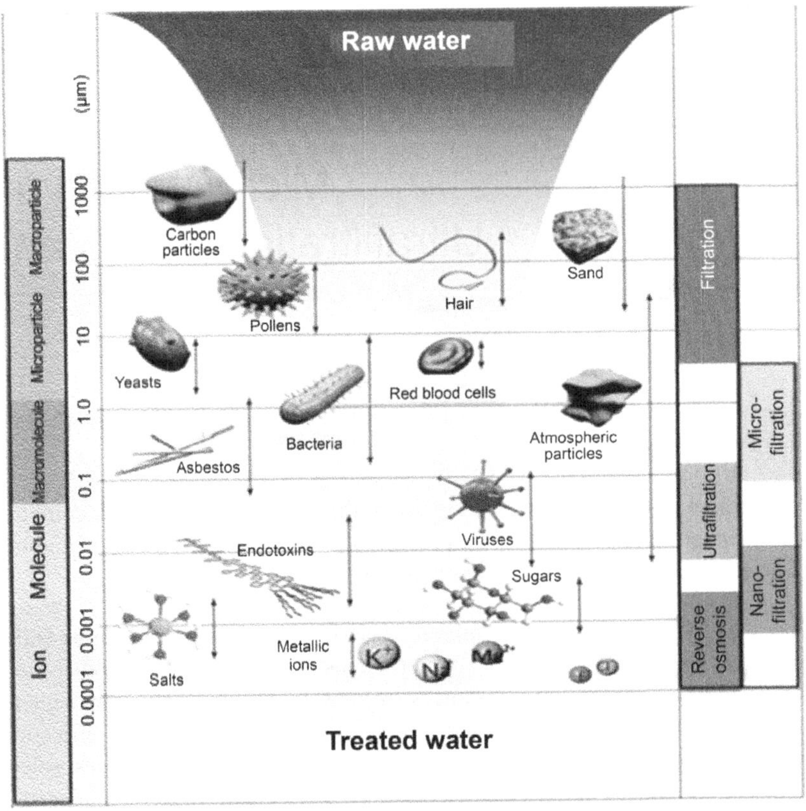

Figure 3.1. *Classification of different species according to size and membrane cut-off point. For a color version of this figure, see www.iste.co.uk/gaid/reuse2.zip*

Membrane clogging is often defined in terms of the type of the mineral or organic substances that are present in the feed stream and can be classified into four categories:

1) particulate/colloidal clogging, when suspended solids and/or colloids clog the pores of a membrane or adhere to its surface;

2) organic clogging, when dissolved reactive organic components or colloids attach to the membrane by adsorption;

3) inorganic clogging, also known as scaling or mineral clogging, caused by the presence of crystallized salts, oxides and hydroxides in the feed solution that tend to precipitate on the membrane surface due to pH change or oxidation;

4) biological clogging, caused when microorganisms, plants, algae or other biological contaminants grow on or in the membrane surface and pores.

Each of these types of clogging can be reversible or irreversible and can permanently compromise membrane performance. All types of clogging can occur simultaneously, and the interaction between them generally increases filtration resistance. Consequently, in every membrane process, clogging problems need to be considered and optimized in order to achieve long-lasting, excellent filtration performance. In addition, an important concept to consider is the threshold flux, which is the flux at or below which a low and almost constant fouling rate occurs, but above which the fouling rate significantly increases.

3.1. Operating principle and mechanisms

MF and UF are two techniques designed for solid/liquid separation, as well as removal of particles and macromolecules. They are generally used for their ability to produce extremely low-turbidity water.

The main suppliers of MF/UF membranes for wastewater applications include Pall/Asahi Kasei, Siemens/Memcor, GE/Zenon and Pentair/Norit.

The main water quality parameters that define MF and UF sizing are suspended solids, turbidity, which reflects colloidal content, and total organic carbon (TOC) concentration. These parameters can be interdependent when colloidal particles are bound to organic matter.

MF and UF are found in fairly wide pore size, ranging from 0.1 to 1 μm. Typically, MF membranes used in wastewater have a pore size between 0.05 and 0.2 μm, while UF membranes used for the same application have a pore size close to 0.01 to 0.03 μm or a MWCO of 80–200 kDa. MF is therefore essentially aimed at retaining particles between 0.1 and 5 μm in size, whereas UF particle sizes are between 0.1 and 0.01 μm.

The cut-off value is the molar mass of the smallest compound that is 90% retained by the membrane. The lower the cut-off value, the greater the membrane's ability to retain small molecules or colloids. Conversely, the higher the cut-off value, the lower the membrane's ability to retain small molecules. For MF, the cut-off is generally expressed in terms of pore size (μm), whereas for UF, it is expressed in terms of molecular weight cut-off (MWCO), which is grams per mole.

The dalton is equivalent to one-twelfth of the mass of a carbon-12 atom, expresses the mass of a hydrogen atom and is equal to 1.66×10^{-27} kg (in other words, 10^{-3}/Avogadro number). Therefore, a protein with a molar mass of 75,000 $g.mol^{-1}$ will have a mass equal to 75,000 Da or 75 kDa.

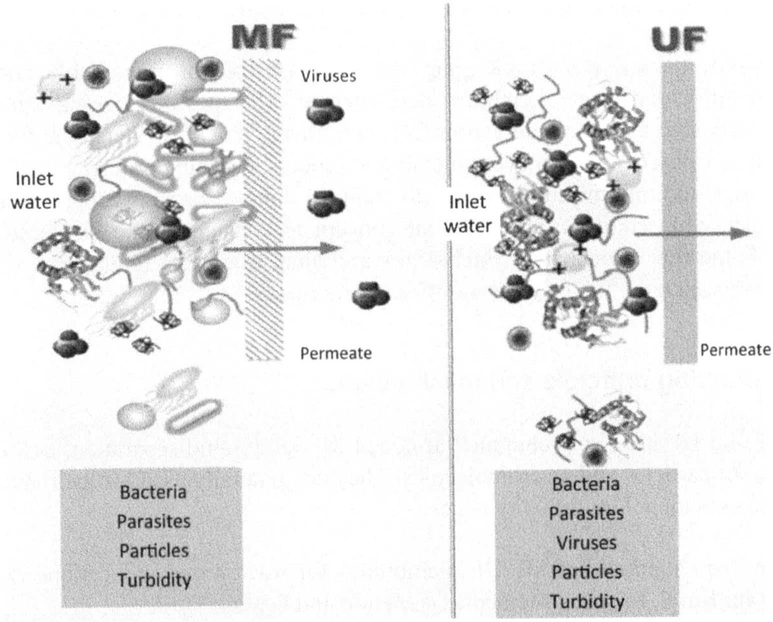

Figure 3.2. *Microorganism cut-off with filtration, microfiltration and ultrafiltration. For a color version of this figure, see www.iste.co.uk/gaid/reuse2.zip*

MF and UF are often distinguished by log virus removal. Taking MS2 phage as a reference, a membrane is considered to be UF when the MS2 removal log is at least 4 log.

UF has two features that make it particularly well suited to water treatment:

– First, pore size ensures the retention of bacteria and viruses, while allowing dissolved salts to pass through, thus combining food safety with the preservation of mineral balance, both of which are essential for drinking water.

– Secondly, UF uses relatively low TMPs compared to NF or reverse osmosis, which has the effect of limiting operating costs associated with energy consumption, with the latter being linked to TMP.

There are a variety of membrane types and modules for UF and MF membranes. Their dimensions and implementation are not standardized. It can sometimes be difficult to change membranes if this has not been taken into account from the outset of the project.

Some membrane manufacturers, however, can offer replacement products with the same dimensions and using the same equipment when renewing membranes at a plant.

The most common type of membrane on the MF/UF water market is hollow fiber, as it is compact and well suited to this type of application (cleaning, frontal or tangential modes, etc.).

These membranes correspond to cylinders of variable length and very small diameter. The internal diameter can vary from 0.3 to 1.5 mm.

There are two types of hollow-fiber membranes: single-bore and multi-bore.

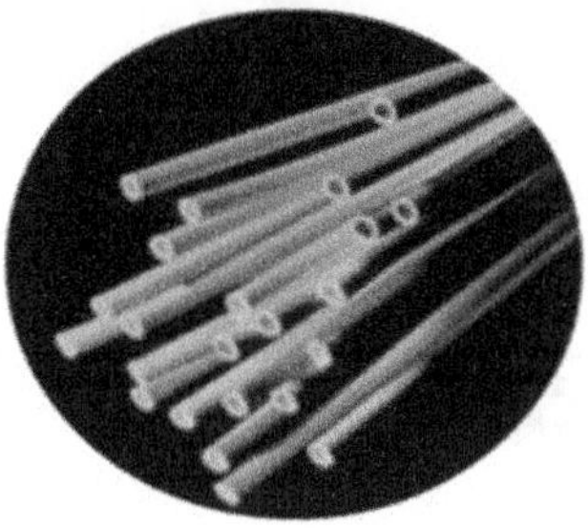	Single-bore hollow fibers These are the fibers offered by most membrane manufacturers. They can be found in pressurized or submerged configurations. They can sometimes be fitted with a support layer for membranes, requiring high mechanical strength. Depending on the type of module, filtration can take place from the inside to the outside of the fibers, or from the outside to the inside.
	Multi-bore hollow fibers In this configuration, each fiber has several internal channels, giving it greater mechanical strength. The material used is hydrophilic polyethersulfone. Filtration only takes place from the inside of the fiber to the outside. This particular type of hollow fiber is currently only mainly offered by one membrane manufacturer.

Table 3.1. *Single-bore and multi-bore hollow-fiber membranes*

These hollow fibers are grouped into bundles of several thousand fibers, and each bundle is placed in a cylindrical module. The ends of the fibers are sealed on both sides in resin, allowing feed water access from inside the fibers.

The feed water therefore penetrates the interior of the fibers and can only pass through tiny pores distributed along the entire length of the fiber, while the particles it carries remain inside the fiber.

The filtered water, or permeate, is isolated from the feed water compartments by the resin used to seal the fibers. The permeate is then collected in a central tube and discharged through the ends of the modules.

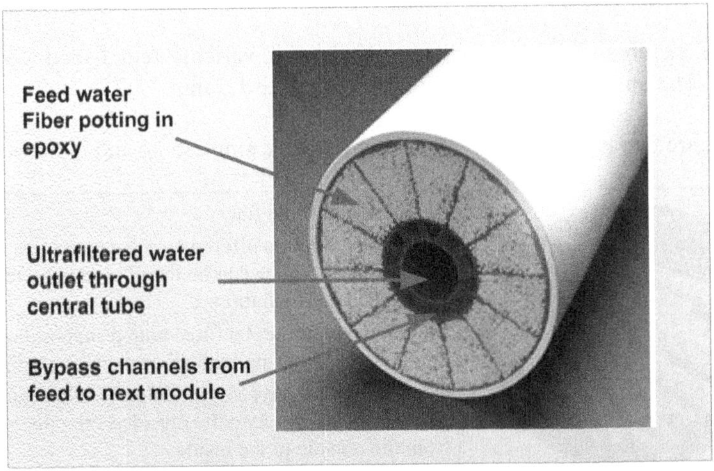

Figure 3.3. *Ultra-filtered water outlet feed points. For a color version of this figure, see www.iste.co.uk/gaid/reuse2.zip*

Hollow fibers are supported in cylindrical housing modules for pressurized membranes, or in more or less open modules for submerged membranes. The fibers are secured by epoxy potting at their ends, which ensures physical separation between the feed and permeate sides of the membranes.

In some cases, pressurized modules contain a central collector that requires seals to be fitted between the feed and the permeate.

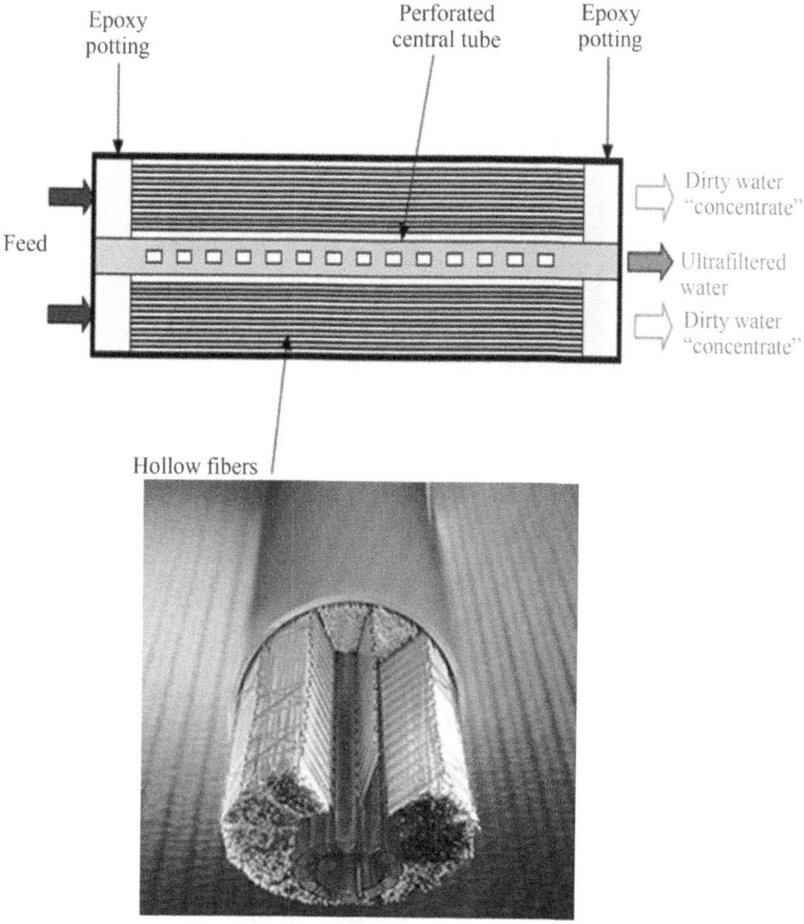

Figure 3.4. *Example of a hollow fiber module in internal/external configuration with central collector. For a color version of this figure, see www.iste.co.uk/gaid/reuse2.zip*

3.1.1. *Pressurized membranes*

Membranes are inserted into housings or pressure tubes, which are installed on a skid and pressurized by a booster pump. The pressure on the feed side of the module or pressure tube provides the pressure gradient required for water to pass through the membrane pores. In this way, a difference in pressure is imposed on both sides of a semi-permeable membrane.

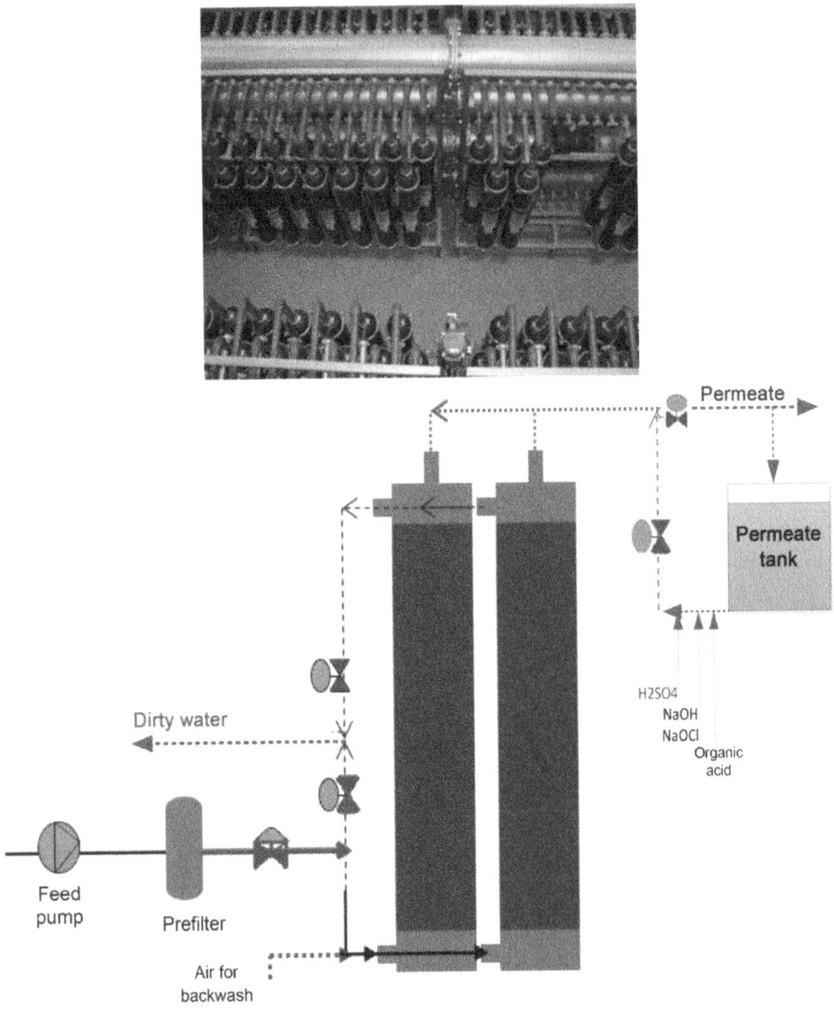

Figure 3.5. *Membranes in a pressurized configuration. For a color version of this figure, see www.iste.co.uk/gaid/reuse2.zip*

3.1.1.1. *Horizontal/vertical configuration*

Depending on the application, pressurized or submerged membranes can be installed in horizontal or vertical configurations.

Vertical pressure configuration Horizontal pressure configuration

Figure 3.6. *Vertical/horizontal configuration for pressurized membranes. For a color version of this figure, see www.iste.co.uk/gaid/reuse2.zip*

Pressurized MF/UF membranes are usually installed in a vertical configuration.

Some membrane suppliers offer a horizontal arrangement in pressure tubes, with several modules per pressure tube. In some cases, this configuration can result in a more compact solution in terms of floor space.

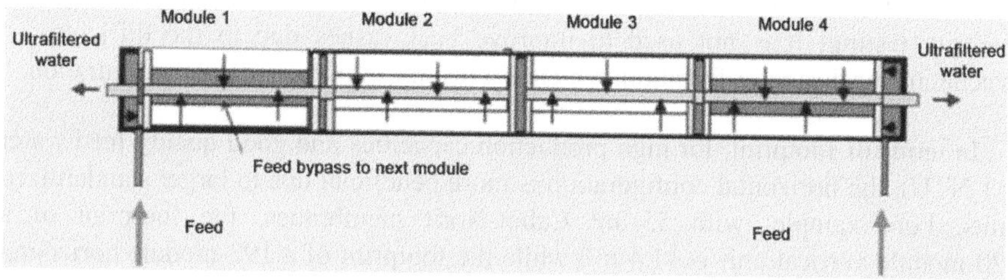

Figure 3.7. *Horizontal filtration mode (Xiga for Cabot-Norit)*

Inside the pressure tube, the membranes operate in parallel. In the case of a four-element tube, the ultrafiltered water is fed and discharged from both sides. The modules have a bypass system surrounding the permeate collector, allowing raw water to be fed to the central modules and dirty backwash water to be discharged. Interconnecting parts between the modules ensure hydraulic operation in the pressure tube.

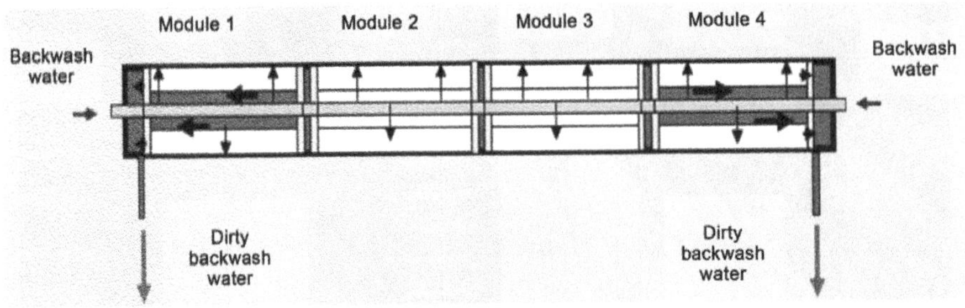

Figure 3.8. *Horizontal backwash filtration mode (Xiga for Cabot-Norit)*

The vertical configuration provides tangential filtration and air backwash conditions (backwash, rinse, air rinse and chemical backwash). Its more robust operation makes it the ideal choice when raw water quality varies throughout the year.

A horizontal configuration is recommended for water qualities below 10 NTU. This means that for wastewater, it is installed behind a lagoon, gravity or mechanical filtration.

In a horizontal configuration, four modules are combined in a pressure tube. This has the disadvantage of making it more difficult to identify a module with broken fibers, as it will be necessary to isolate this tube and test the modules one by one to determine which one is faulty. In addition, the use of air is rather restricted to integrity testing; it is not used to improve backwashes due to the difficulty of evacuating air during re-filtration, which is a result of this horizontal configuration.

In terms of footprint, for high production capacities and good quality feed water (<1 NTU), the horizontal configuration is more beneficial due to larger standardized units. For example, with 55 m^2 Cabot-Norit membranes, the footprint of a 120-module vertical unit is 41.4 m^2, while the footprint of a 192-module horizontal unit is 16.7 m^2.

Size is one parameter considered by the manufacturer, but operating conditions are often the selective parameter. For example, vertical configuration membranes are typically chosen for small installations, or when powdered activated carbon (AC) needs to be added, if tangential filtration is to be implemented, if turbidity is variable (notably in direct raw water feed), or for the addition of a coagulant and integrity tests, which are easier to implement than for horizontal modules.

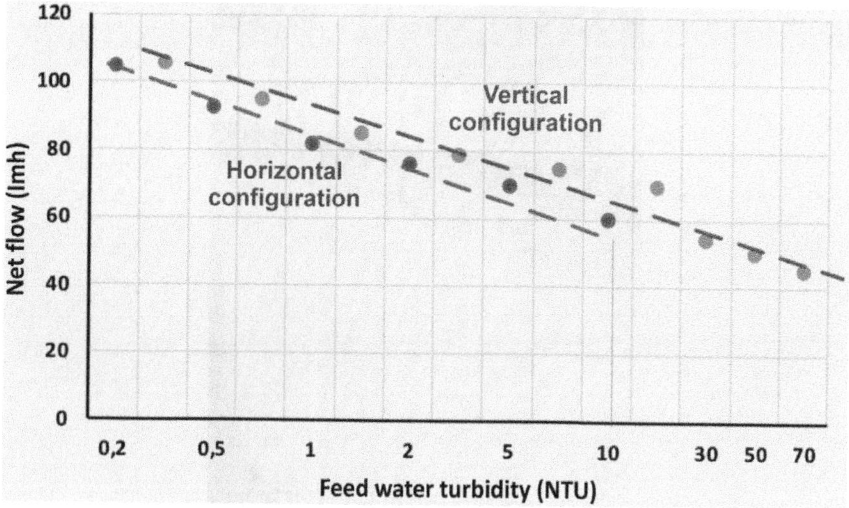

Figure 3.9. *Recommended net flux for both configurations as a function of turbidity*

The vertical configuration is therefore most often used. It allows for more complex backwash sequences if necessary, with both air and water phases, and therefore for the treatment of more heavily loaded water.

In addition, as the vertical modules are individually connected to the collector and the horizontal modules are loaded into pressure tubes, there is a higher permissible operating pressure for horizontal configurations.

3.1.2. *Submerged membranes*

Submerged MF membranes are available in vertical configurations, such as the vertical modules for Evoqua's CMF-S (ex-Memcor).

The membranes are submerged in a tank (made of steel, high density polyethylene (HDPE), coated concrete, etc.) called a cell, and filtration takes place by drawing water through the membrane using a pump placed at the permeate outlet of the membrane cell.

The CMF-S® cell uses hollow-fiber membranes. Feed water flows into a cell containing submerged membrane modules and is drawn through the membrane fibers in an external-internal filtration direction. The microfiltered water exits at the top of the module, and is then led to a central collector.

Figure 3.10. *Membranes in a submerged configuration. For a color version of this figure, see www.iste.co.uk/gaid/reuse2.zip*

CMF-S® systems can be fitted with both polypropylene (PP) and polyvinylidene difluoride (PVDF) membranes.

A backwash system uses a combination of air and water backwash to remove the build-up of solids deposited on the membrane's outer surface. Backwash frequencies range from 15 to 60 min.

Chemical cleaning is programmed to remove organics, minerals and biofilms that have not been removed by hydraulic and chemical backwashing.

The CMF-S system is designed with an automated integrity test, called a pressure decay test (PDT).

Washing efficiency is obtained using the following equation:

$$\text{Washing efficiency} = [1 - \left(\frac{R_3 - R_1}{R_2 - R_1}\right) \times 100\%].$$

where:

– R_1 is the resistance calculated after a previous wash ($m^{-1} \times 10^{12}$);

– R_2 is the resistance calculated before a wash, after an operating time corresponding to a volume of filtered water ($m^{-1} \times 10^{12}$);

– R_3 is the resistance calculated after this wash ($m^{-1} \times 10^{12}$).

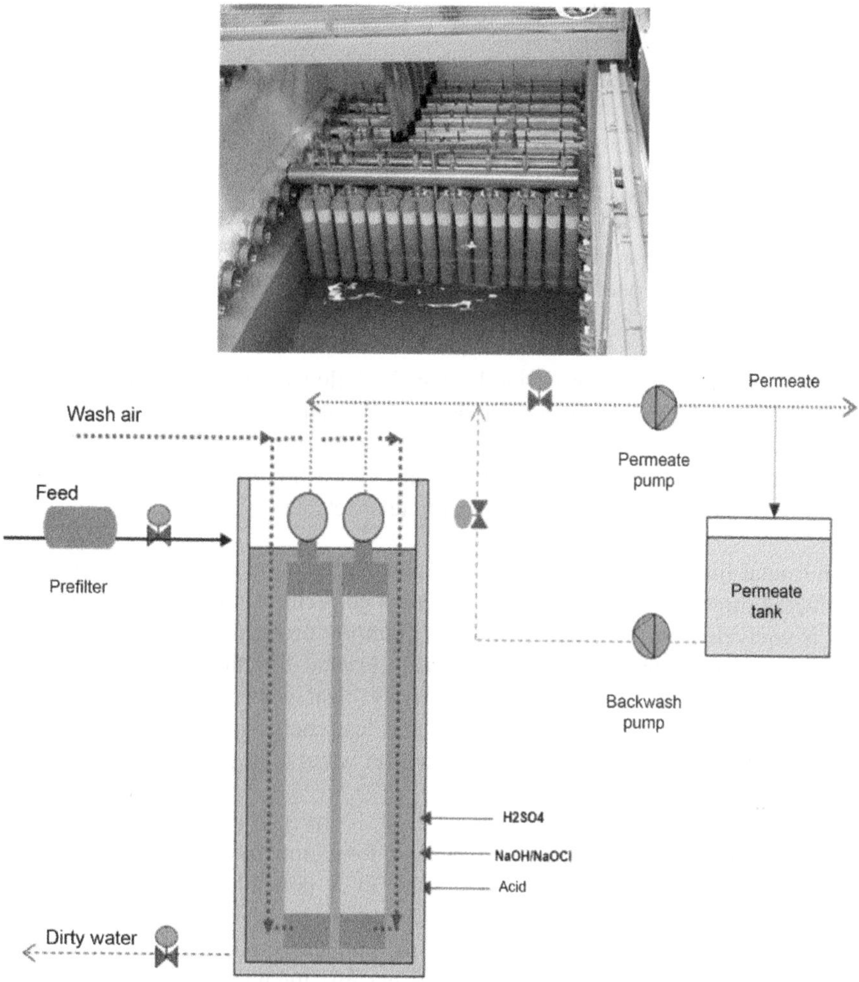

Figure 3.11. *Membranes in a submerged configuration. For a color version of this figure, see www.iste.co.uk/gaid/reuse2.zip*

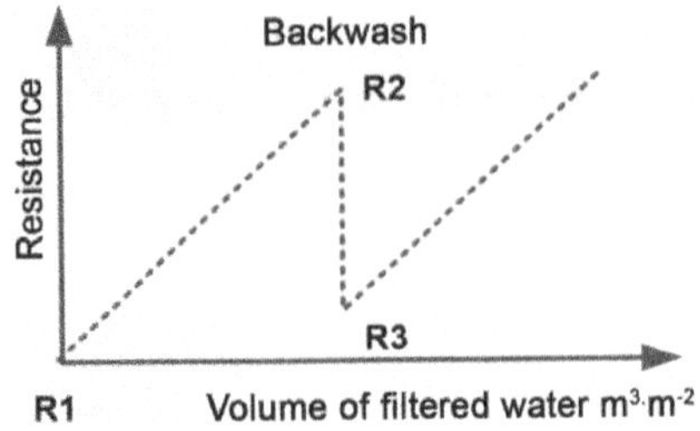

Figure 3.12. *Washing efficiency based on resistance to fouling*

3.1.3. *Filtration mode: frontal/tangential*

Depending on the quality of the water to be treated and the objectives to be achieved, UF can be operated in two flow modes: tangential filtration or frontal filtration.

When filtering in tangential mode, the fluid flows parallel to the membrane, creating shear that limits the accumulation of material. This mode of operation enables continuous water production. There is necessarily an inlet for the feed and two outlets: one for the permeate and the other for a flux of water, which does not pass through the membrane, called concentrate or retentate.

In frontal filtration, the feed flow is perpendicular to the membrane and particles are filtered directly by the membrane. There is no recirculation pump, and the feed flow is equivalent to the product flow. As filtration proceeds, particles accumulate and form a deposit that clogs the membrane. Frontal filtration is more commonly used for hollow-fiber membranes in relatively clear water applications (drinking water, process water, tertiary wastewater). It is less energy-intensive than tangential filtration, which requires much larger volumes of water to be pumped.

In tangential filtration, a tangential flow limits particle accumulation. This process helps to limit clogging and deposit formation by creating shear forces. However, recirculation of the flow results in a potentially higher number of membranes and additional costs due to recirculation energy. Tangential filtration is often used with hollow-fiber membranes for applications involving contaminated water or the addition of powdered AC. It ensures membrane cleaning and, to a certain extent, avoids deposits on the membrane surface.

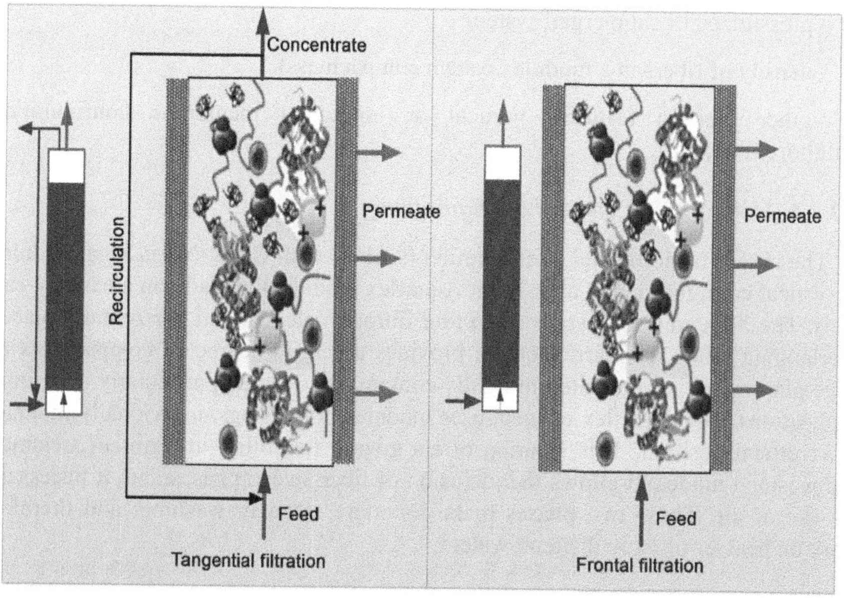

Figure 3.13. *Tangential and frontal filtration. For a color version of this figure, see www.iste.co.uk/gaid/reuse2.zip*

3.2. Sizing parameters

3.2.1. *Membrane selection*

The first sizing phase involves choosing the type of membrane. Choosing a membrane is a complex process, as it not only depends on the purpose of the treatment, but also on the various physicochemical and biological conditions that prevail when treating water or a solution by membrane filtration.

The technical parameters to be considered when choosing a membrane are:

– quality of water to be treated;

– treatment objectives;

– MF or UF;

– membrane type: hollow fiber, tubular, spiral wound, etc. Usually, for UF/MF in drinking water, hollow fibers are used;

– type of filtration (for hollow fibers): internal-external or external-internal;

– footprint;

– pressurized or submerged system;

– density of fibers in a module (system compactness);

– other specific features: vertical or horizontal membrane configuration, multibore fiber, etc.

3.2.1.1. *Horizontal or vertical configuration*

The choice of horizontal configuration (such as with Cabot-Norit Xiga modules) or vertical configuration (Cabot-Norit Aquaflex modules) is made on a case-by-case basis. The Xiga concept installs up to four filtration elements in horizontally placed, overlapping tubes. This arrangement provides a certain degree of compactness for large plants, and is well suited to lightly contaminated water, particularly in refining applications. The Aquaflex concept uses modules in housings, individually mounted in a vertical position. This solution offers greater flexibility in terms of treatment and washing modes. It allows the inclusion of fiber sweep phases and, if necessary, the use of air. These two phases make for more efficient washing, and therefore work on heavier or more difficult water.

3.2.1.2. *Sizing flux*

Darcy's law applied to water treatment is used to translate the proportional relationship between permeate flux J and TMP:

$$J = \frac{TMP}{\mu s\ Rm} = Lp - TMP$$

where Rm is the membrane resistance, Lp is the membrane permeability and µs is the viscosity at water temperature.

In the case of water containing a solute retained by the membrane, such as salts or organic substances, transfer resistance is added via an osmotic pressure difference ΔΠ opposed to the pressure gradient TMP, due to the difference in concentration on either side of the membrane. Darcy's modified law can then be used to describe the flow according to the following relationship:

$$J = \frac{TMP - \Delta\Pi}{\mu s\ Rm}$$

Permeate flux no longer evolves linearly with TMP. The difference in osmotic pressure is neglected in MF and UF, as the compounds retained are large in size, inducing little osmotic pressure. Nevertheless, the phenomenon of concentration polarization may occur.

Organic substances and particles can clog the membrane more or less irreversibly, due to specific physical and/or chemical interactions between them and the membrane, resulting in additional transfer resistance.

These interactions are generally attributed to one or more of the following mechanisms: deposit formation, pore blocking and adsorption.

The relative predominance of each of these phenomena may depend as much on the membrane (material, cut-off threshold) as on the nature of the substances, the characteristics of the medium (pH, temperature, ionic strength, etc.) or the operating conditions.

3.2.1.3. *Instantaneous and net flux*

Permeate flux corresponds to the flow rate per m^2 of membrane. It is usually expressed in liters per hour per m^2, known as lmh, but can also be expressed in meters per day ($m.day^{-1}$). Just as we distinguish between net flow rate and filtration flow rate (or instantaneous flow rate), we distinguish between net flux and instantaneous flux in filtration.

The net flux at a given temperature (T°) is determined by the following formula:

$$J_{NET}(T°) = \frac{Q_{net}(T°) \times 10^3}{S_f}$$

with:

– J_{NET}: net flux at temperature T° ($L.m^{-2}.h^{-1}$);

– Q_{net}: net filtration flow rate at temperature T° ($m^3.h^{-1}$), daily flow rate produced (excluding UF wash volumes) per hour of membrane plant operation;

– S_f: total filtration area (m^2).

As temperature has an influence on flux, it is important to relate flux to a temperature-related reference value. For example, between sizing for water at 12°C and sizing for water at 6°C, flux and production flow rate will be around 16% lower at 6°C than at 12°C, for the same operating pressure.

Submerged membranes operate at lower flow rates, as they rely on the application of a partial vacuum on the permeate side, and therefore operate at lower pressures.

UF feed water	Maximum turbidity (NTU)	Maximum instantaneous flow at 20°C ($L.m^{-2}.h^{-1}$)	Estimated hydraulic efficiency (%)
Treated wastewater	>2.0; ≤4.0	80	90.5–91.6
	>4.0; ≤10.0	70	85.5
	>10	50	75

Table 3.2. *Examples of flux size and chemical backwash frequencies for pressurized membranes*

MF/UF design flux values in all plants range from 20 to 60 lmh, depending on water quality and temperature.

The application of high fluxes for given conditions and for a specific membrane configuration can quickly lead to clogging and the need for more frequent cleanings. TMP ranges from 0.3 to 0.6 bar for MF membranes, and from 0.8 to 1.2 bar for UF membranes.

Conversion rates vary between 80% and 95% depending on the membrane and its age, reflecting the low water losses.

3.2.1.4. *Principle for calculating membrane surface area and water loss*

MF and UF are intermittent processes, with frequent skid or cell shutdowns for washing, chemical washing and integrity testing. This creates production and downtimes for each skid or cell.

When the specifications call for a constant flow to the membrane plant, the membranes operate with a variable flux rate, a flux rate in N when all skids are filtering, and a flux rate in N-1 when one of them stops production. The flux size for the membrane plant must therefore be based on the N-1 flux, in the same way as for gravity filtration.

When the workshop is designed for a nominal daily output that considers skid downtime, the skids are sized to produce the required daily output (including water losses due to washing), taking downtime into account.

A buffer tank eases the flow between the upstream process and the requirements downstream of the membrane plant.

The membrane surface is determined from:

– the duration of non-production phases (or downtime) for each skid;

– water losses due to MF/UF washing.

The number of daily production hours for the membrane workshop is defined, as well as its operating mode (fixed or variable flux). For each skid, the number of hours linked to backwash (BW), chemical backwash (chlorine, acid or basic) and integrity test cycles is determined, as the skid is shut down during these maintenance periods.

As each skid undergoes the same downtime, this implies that the non-production time of each skid is equivalent to the plant's total non-production time.

As there are few chemical cleanings in place, downtimes may not be taken into account.

3.2.1.5. *Water losses*

The water loss rate is linked to backwash frequencies, which depend on the flux applied, membrane configuration and water quality (turbidity and TSS), as these parameters influence the permeability drop (Espinasse et al. 2002; Farahbakhsh et al. 2004).

Backwashing is known to mainly remove reversible fouling, while additional periodic chemical cleaning is required to remove irreversible fouling (Fan et al. 2001; Farahbakhsh et al. 2004; Goosen et al. 2004). Chemical cleaning is normally carried out without removing the membranes (hence "cleaning in place" (CIP)), and cleanings often combine backwashing with chemical cleaning (a "chemically enhanced backwash" (CEB)). Cleaning frequency and chemical dosing requirements mean that CIP frequency for MF and UF membranes should, in most cases, be carried out once or twice a month.

Sodium hypochlorite (NaOCl) is widely used for CIP and CEB chemical cleaning, and can be combined with sodium hydroxide (NaOH) to improve cleaning performance, provided the membranes can tolerate high pH levels. Chelating acids such as citric and oxalic acids are used to complement hypochlorite cleaning, when metal precipitation is observed.

During backwash sequences, part of the permeate and, in some cases, part of the feed water, is consumed and discharged to the sewer. Calculating the daily volume of permeate used for backwashing operations gives the water losses for the membranes workshop. The workshop is therefore sized to take account of the volume consumed. For some membranes, a volume of feed water is also consumed during washing cycles (fiber sweeping, cell emptying, etc.) and must therefore be considered.

For each type of sequence (BW, CEB integrity test and CIP), the following calculations are performed:

– number of sequences per day (per skid or cell and total);

– duration of each sequence;

– volume of permeate consumed by each sequence;

– volume of feed water consumed by each sequence.

For each skid, we deduce:

– workshop non-production time, corresponding to the sum of the number of sequences × sequence time;

– the total volume of permeate consumed by each skid or cell, which corresponds to the sum of the number of sequences × volume of permeate in the sequence;

– the total volume of feed water consumed by each skid, which is the sum of the number of sequences × sequence feed water volume.

The volumes consumed are multiplied by the total number of skids or cells, giving the overall water losses.

This gives the total daily production volume, which takes the plant's internal permeate requirements into account. The workshop's actual total production time takes into account production downtimes when a skid or cell is in the non-production phase.

This information is used to deduce the total membrane surface area required to produce the required permeate flow rate based on a flux defined at the outset (depending on the quality of the water to be treated).

Membrane clogging is the main handicap to membrane systems, resulting from the accumulation of material on the membrane surface. The operation of a membrane filtration process is conditioned by clogging, which is the phenomenon that opposes filtration.

When a system operates at constant TMP, clogging results in a decrease in permeation flux over time. In constant-flow operations, such as water treatment, the direct consequence is an increase in pressure.

The concept of critical flux is defined as the phase transition between reversible clogging (concentration polarization) and irreversible clogging (deposit formation). The method for determining this transition is based on the use of pressure, the application of flow steps and the critical filtered volume, defined as the filtered volume at which clogging becomes irreversible. It appears that charged colloidal particles do not adsorb to the membrane surface directly and irreversibly, but only after a filtration or critical time. The critical filtered volume is determined as the junction between the tangents of the resistance as a function of the net filtered volume curves.

The damaging role of internal pore fouling is counted as irreversible fouling, while surface fouling acts as a kind of barrier preventing access to the pores. It has been shown that non-electrostatic organic substance/membrane interactions are the dominant mechanism of irreversible fouling.

In some cases, several cycles of chemical cleaning have produced no change in membrane permeability. Progressive irreversible chemical fouling therefore leads to a loss of productivity. Irreversible fouling can be accompanied by a progressive loss of mechanical strength in the membrane.

Clogging is caused by an accumulation of material at the interface between the membrane and the water, resulting in an increase in TMP during constant-flow filtration. Regardless of the mechanism involved, clogging can either be reversible or irreversible. Reversible clogging occurs when, after hydraulic or chemical washing, the clogging deposit is eliminated. This results in very good recovery of permeability and TMP, which return to their initial values when filtration is resumed. If hydraulic or chemical washing does not restore permeability, this is known as irreversible clogging. Consequently, when filtration resumes after the washing procedure, the TMP will be higher than its initial value.

This clogging can be:

– particulate (deposition of particles on the membrane surface);

– biological (deposition of a biofilm on the membrane surface);

– organic (dissolved organic carbon (DOC), color adsorbed on the membrane);

– inorganic (calcium carbonate, iron, aluminum, manganese, etc.).

Material build-up requires frequent hydraulic and chemical cleaning, which affects membrane life, water losses and operating costs.

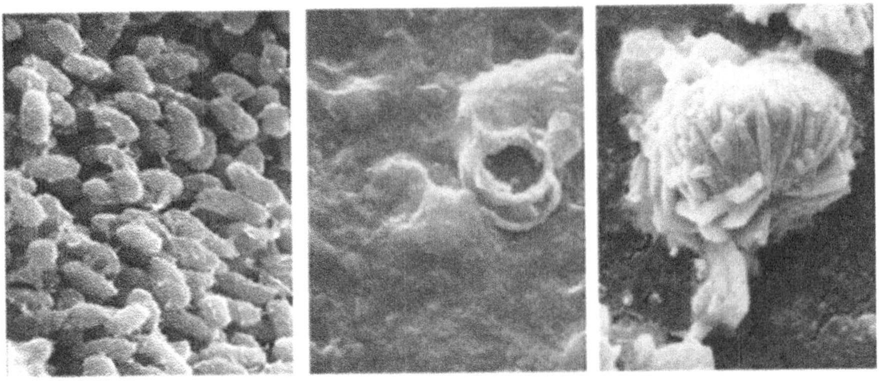

Figure 3.14. *Different types of clogging*

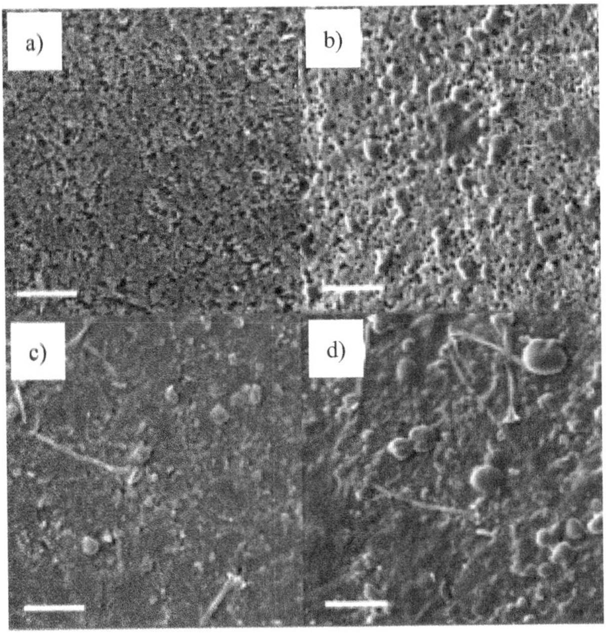

Figure 3.15. *Development of clogging on an MF membrane after several backwash cycles. a) New membrane; b)–d) membrane after backwash (Gupta 2021)*

The physicochemical properties of membranes determine whether they are hydrophobic or hydrophilic. Hydrophilic membranes are characterized by the presence of active groups that form hydrogen bonds with water, whereas hydrophobic membranes have little attraction to water molecules, which tends to favor clogging with hydrophobic particles and substances. Other parameters such as pore size, pore distribution and pore geometry influence clogging mechanisms, such as pore blockage and the accumulation of material on the surface. The same applies to particles and substances that are present in the water, whose physicochemical properties may or may not be involved in clogging.

Surface filtration favors the accumulation of matter (TSS and high-molecular-weight molecules) on the membrane surface. As the water is filtered, the pressure required to maintain the same flow rate across the membrane increases.

Many parameters make the behavior of a UF or MF membrane more complex:

– formation of a cake on the surface of the membranes, which introduces depth filtration behavior;

– since the membrane surface is electrically charged, electrical repulsions are involved in the separation mechanisms;

– molecules with molecular weights lower than the membrane's cut-off power, such as certain organic materials, can sometimes adsorb to the membrane surface.

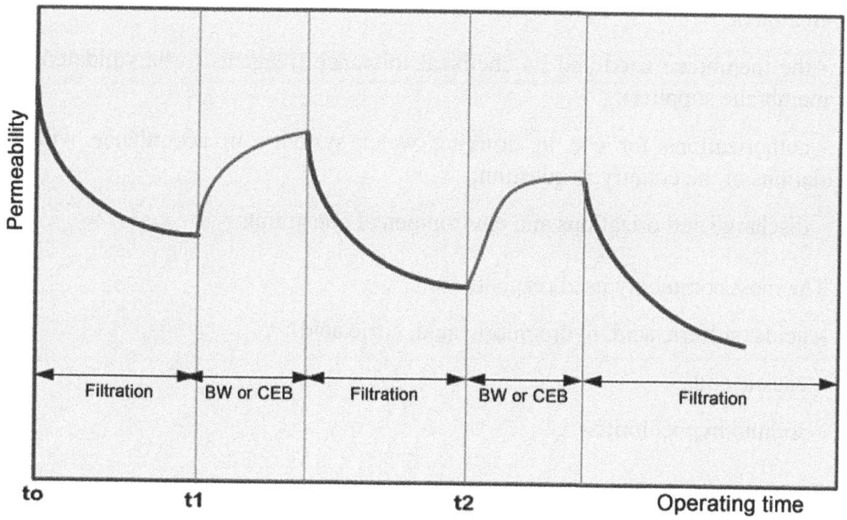

Figure 3.16. *Sequential diagram of permeability recovery after clogging*

3.2.2. *Cleaning in place*

Given the performance of hydraulic filtration, almost all researchers have been interested in controlling clogging. Periodic backwashing could extend UF operating life, reduce the frequency of chemical cleaning and save energy during long-term operation. Therefore, to sustain high MF/UF performance in wastewater treatment in long-term operation, appropriate physical and chemical cleaning of membranes is necessary.

CIP is a corrective cleaning procedure designed to restore membrane permeability. The proportion of clogging that cannot be recovered by backwashing is said to be totally irreversible. It is generally attributed to adsorption and/or precipitation phenomena.

When a membrane's capacities drift too largely, which is a limit defined by the manufacturer, more thorough cleaning must be carried out in order to restore the membrane's permeability (this is complementary to hydraulic and chemical backwashing).

The choice of reagents depends on:

– the type of clogging encountered, and therefore the water to be treated and the pre-treatment;

– the membrane used and its chemical tolerance (reagents to be validated with the membrane supplier);

– authorizations for use in drinking water systems, in accordance with the regulations of the country in question;

– discharge authorizations and environmental constraints.

The most commonly used reagents are:

– acids (sulfuric acid, hydrochloric acid, citric acid);

– caustic soda;

– sodium hypochlorite.

In agreement with the membrane supplier, other products are listed, such as detergents, oxalic acid or Hydrex 4920, or products specifically developed for membrane cleaning (Veolia-Hydrex).

Cocktails of products such as soda + sodium hypochlorite, citric acid + sulfuric acid and detergent + soda are also often recommended.

3.2.2.1. *CIP frequency and conditions*

CIP is less frequent than chemical backwashing. They require more severe conditions, with longer soaking and recirculation times (several hours), as well as the use of heated solution. On average, these cleaning operations take place two to four times a year. However, this frequency can be increased to monthly if the quality of the feed water to the membranes requires it.

3.3. Microfiltration and ultrafiltration applied to wastewater treatment

In wastewater treatment, MF is located differently. MF is placed behind the secondary clarifier or behind a first filtration stage. UF is more generally placed behind a preliminary filtration stage in order to reduce the concentration of suspended solids as much as possible.

UF separates particles or molecules larger than 0.01 µm at operating pressures ranging from 1 to 3 bar. UF removes bacteria, viruses and proteins. UF is also presented as a system for removing all organic substances with a molecular weight greater than 150 kDa.

3.3.1. *Turbidity and TSS*

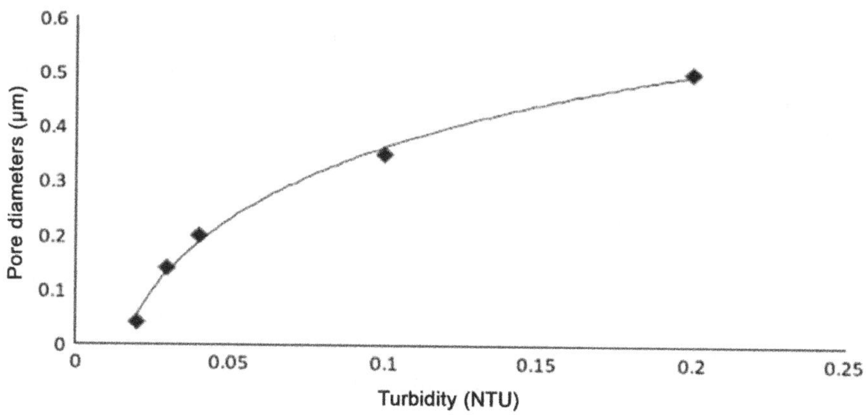

Figure 3.17. *Turbidity of filtered water as a function of MF/UF pore diameter*

In wastewater reuse, UF is used for frontal filtration of water with an average turbidity ≤2 NTU, in other words, water that has already undergone material filtration.

MF and UF performance usually delivers turbidities <0.5 NTU, whatever the feed water quality. Turbidity reduction is linked to membrane pore size, with the 0.2 μm membrane delivering the highest filtrate turbidity and the 0.02 μm membrane delivering the lowest (Figure 3.17).

High filtrate turbidity leads to early RO membrane clogging.

3.3.2. *Dissolved organic carbon*

For design purposes, it is assumed that MF and UF membranes do not retain dissolved organic matter (DOM), even though around 10% of DOC is retained by UF membranes through adsorption, and contributes to membrane clogging. For waters with more than 2 mg.L^{-1} of DOC, this adsorption will lead to a loss in permeability, requiring frequent chemical backwashing with soda (and even chemical cleaning for very difficult waters). In addition, the UF removal efficiency on chemical oxygen demand (COD) and DOC concentrations is around 5% to 20% higher than that of MF.

MF and UF have no minor effects on the removal of residual nutrients such as phosphorus, nitrates and ammonium. Generally speaking, effluent from wastewater treatment plants still contains high levels of suspended solids (TSS) and natural organic matter (NOM), which can facilitate the formation of fouling on the MF/UF membrane. In fact, DOM cannot be effectively removed by MF or UF systems, but conversely, it can be the main cause of fouling formation on the membrane, ultimately leading to shortened membrane life, reduced throughput, increased TMP and higher energy consumption.

3.3.3. *Energy consumption*

Figure 3.18 illustrates specific energy consumption as a function of flux applied for MF/UF. This consumption is different for submerged membranes, where filtration is achieved by sucking water through the fibers, and pressurized membranes, which involve pushing water through the fibers.

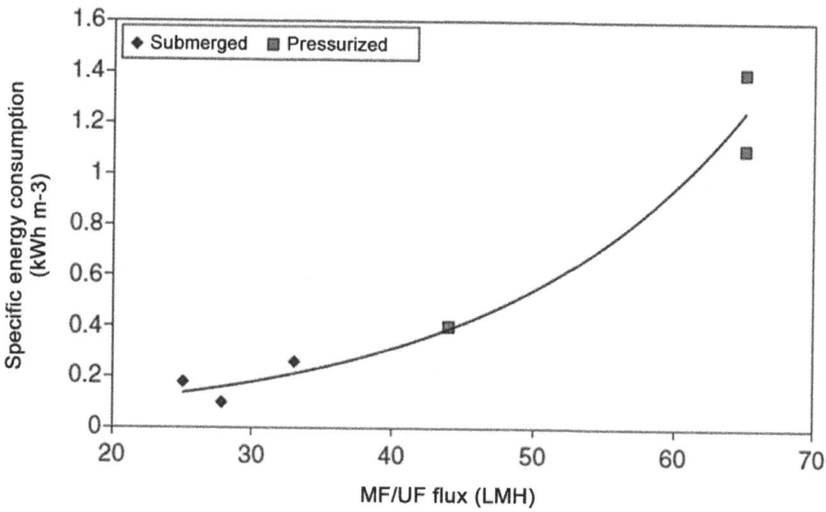

Figure 3.18. *Specific energy consumption by MF/UF flux*

3.4. Improved secondary treatment: membrane bioreactors

In the wastewater industry, treatment plants focus on primary and secondary treatment, which perform the bulk of the removal of dissolved solid and organic substances for discharge into the environment. Primary and secondary treatment generally render wastewater clean enough to be discharged into the environment.

Secondary treatment with submerged membranes can significantly improve the microbiological quality of treated water for parameters such as suspended solids and particulate organic substances.

Conventional activated sludge processes are widely and predominantly used due to their robustness, range of action and cost. However, one of the main weaknesses of this type of treatment is separation by settling. The efficiency of this stage depends on good biological functioning. The appearance of filamentous microorganisms, for example, leads to the sludge overflowing and rendering it unsuitable for settling. The resulting biomass/treated water separation then becomes mediocre. Water produced in this way cannot consistently meet the required discharge standards. Nevertheless, there are ways of controlling these drawbacks by monitoring the biological mass and characteristics of the raw water.

Like chemical and physical processes, biological processes can also be combined with a membrane system, due to its benefit in removing microorganisms. Alternatively, the membrane bioreactor (MBR), a combined system, mainly consists of a conventional activated sludge process and a submerged or external MF/UF membrane process. The biological process, which is mainly used for the reduction of organic matter in wastewater, generally comprises an aeration tank used for biological degradation and a secondary clarifier (sedimentation tank), where the sludge is separated from the treated effluent. The effect of the membrane is to increase the concentration in the biological reactor, retain the particulate phase in the bioreactor and allow the permeate to pass on to the next process or to be rejected and/or reused. In MBRs, biodegradation and membrane separation are carried out simultaneously, maximizing wastewater treatment efficiency. MBRs work best in processes involving efficient biodegradation, a small footprint, smaller sludge production, and good resistance to hydraulic and pollutant loads.

In the case of wastewater reuse, the treated water requires tertiary treatment. This additional step can be avoided by using a membrane separation technique (Figure 3.19).

An MBR plant is a potential alternative to the traditional wastewater treatment plant. It combines a biological treatment process with membrane filtration. This method of separation has replaced separation by settling.

Membranes can be mineral or organic based. Membranes made from mineral materials offer superior filtration performance compared with organic membranes, thanks to their excellent chemical resistance, cleaning flexibility and resistance to clogging. However, their high cost restricts them from being used for municipal applications.

Organic membranes made from polyacrylonitrile (PAN), polyethylsulfone (PES), polysulfone (PS), polytetrafluoroethylene (PTFE) and PVDF are used in wastewater treatment plants. Around 65% of MBR modules on the market are PVDF based.

There are two types of bioreactor: external-loop MBRs and submerged MBRs. External-loop MBRs use tubular organic or inorganic (ceramic) membranes, placed in external recirculation loops. Biomass circulates at high speed in these loops, limiting membrane clogging and reducing the frequency of chemical washing. However, circulation is energy intensive, resulting in high energy costs (10–50 kWh/m^3), making this application too costly to operate.

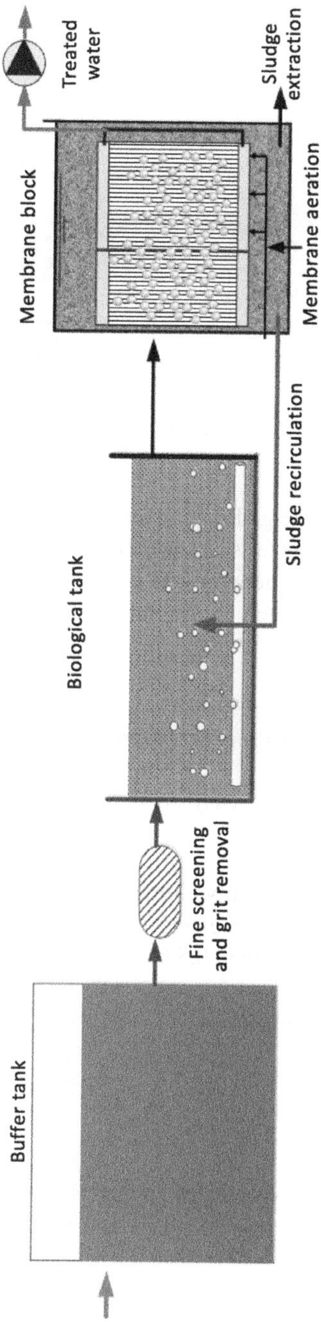

Figure 3.19. *Wastewater treatment with membrane process*

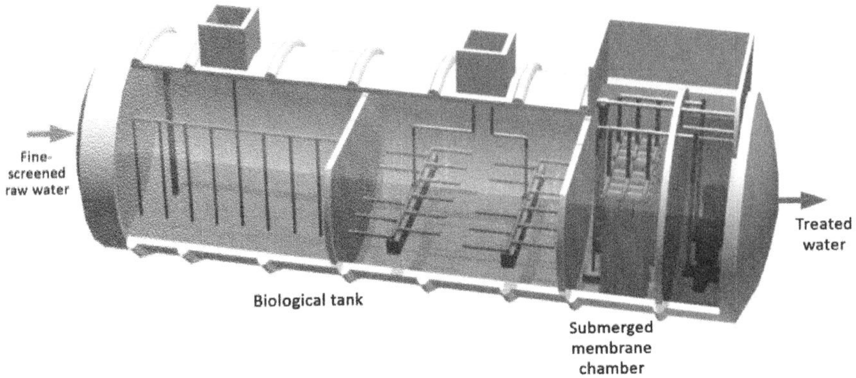

Figure 3.20. *Compact solution with membrane process*

In MBR processes, the submerged membranes can be directly integrated into the biological basins, and the permeate is drawn off by suction. The work is therefore carried out under vacuum, and generally does not exceed 0.5 bar. As a result, the energy expenditure associated with submerged MBRs does not exceed 0.2–0.4 kWh.m^{-3}, which explains why they are more commonly used by municipalities.

There are also lateral-flow MBRs with flat organic membranes.

Hollow-fiber membranes allow easy modulation in their configuration, thus optimizing process performance.

Treated water (the permeate) is only extracted from the bottom of the cassettes, creating a suction effect. Filtration takes place from the outside to the inside of the fibers, under a slight pressure (less than 0.5 bar) known as TMP.

Hydraulic performance is maintained by:

– "gentle" operating conditions (low pressure);

– sweeping the surface of the fibers with air sent in situ to limit material deposition on the fibers;

– short, regular backwashing sequences, in which treated water is re-injected into the fibers in the opposite direction to filtration, keeping filtration performance constant.

TMP and permeability are among the most important critical factors. TMP is the driving force behind filtration through the membrane, while permeability is the ratio of permeate flux to TMP.

MBRs can be operated by maintaining a constant TMP with variable permeate flux, or by maintaining a constant permeate flux with variable TMP. The rate of membrane fouling is exponential to flux, particularly at values above the critical flux, which is defined as the flux at which fouling cannot be controlled by physical cleaning methods.

Cassette operation requires four automatic operating sequences:

– Filtration: permeate extraction pumps draw filtered water through the membranes and return it to the permeate tank (filtration time: 8–12 min).

– Backwashing: permeate pumps draw filtered water from the permeate tank and pump it back into the membranes (approximately six backwashes per hour).

– Degassing: the permeate pumps draw the filtered water into the permeate tank and flood the tank to evacuate the air trapped in the filtered-water pipes. Filtration partly creates a vacuum of the permeate circuit, which encourages degassing (around three degassings per hour).

– Relaxation: filtration is temporarily stopped (only membrane aeration is maintained).

MBR membranes belong to the MF category, and generally have a cut-off between 0.04 and 0.1 μm. According to some suppliers, the membrane offered borders on UF.

3.4.1. *Suppliers of microfiltration membranes used in MBRs*

The water treatment market is full of them:

– organic fiber membranes: Zénon, Evoqua, Puron, Mitsubishi, LG, Litré, Fibracast and Memstar;

– ceramic membranes: Kubota, Alsys, Tami, TFT, etc.;

– flat organic membranes: Alfa Laval, Toray, etc.;

– external membranes: Orelis, X-Flow (Pentair).

The non-drinking application of various MBR systems for water reuse is the subject of numerous references. The water produced has been shown to meet the directives of USEPA and WHO reuse applications, such as toilet flushing, irrigation, car washing, lake recharge and groundwater recharging. MBR treatment has been shown to remove different types of microorganisms more effectively than conventional secondary treatment. For example, the log reduction value (LRV) for all organisms, including enteric viruses, is higher with MBR (4.0–>6.0) than with conventional secondary treatment (1.5–4.0) in municipal wastewater treatment. For fecal coliforms and MS2 coliphages, the ranges are 5.5–6.5 and 2.5–3.5, respectively. Membrane size exclusion is probably the dominant removal mechanism for bacteria and viruses in MBRs with integral membranes. Other removal mechanisms, such as adsorption to biomass, blockage and pore constriction, can also play a major role.

In a MBR, suspended solids are stopped directly by the membrane.

MBR floor areas are much smaller than those of conventional systems, due to the higher organic load and greater depth of the biological tanks. Membrane fouling is one of the main drawbacks of these systems, due to its negative influence on flux reduction, membrane service life, backwash frequency and energy costs. As a rule, membrane fouling is attributed to both membrane pore clogging and sludge cake deposition on the membranes, which is generally the predominant component of fouling.

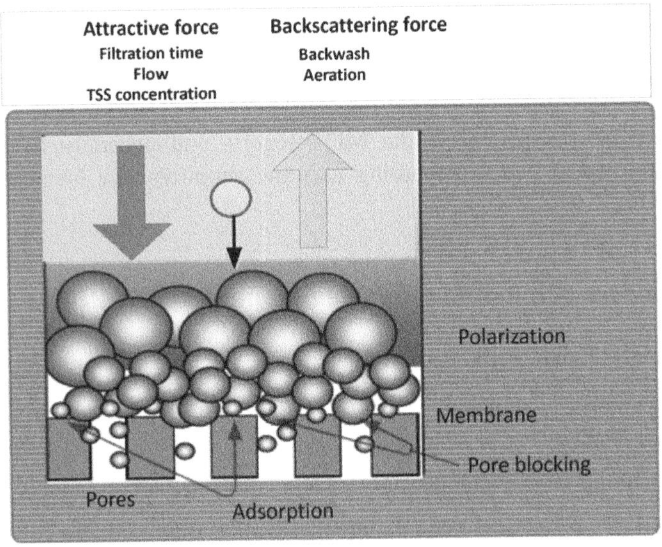

Figure 3.21. *Clogging mechanisms of MBR membranes*

The operation of the MBR process therefore requires careful monitoring due to the potential clogging of the membranes, which requires frequent cleaning and washing with chemicals (chlorine, acid, detergents). Sludge is a complex medium made up of bacterial flocs, colloids and dissolved mineral and organic substances. These two fractions are both involved in membrane clogging mechanisms. Filtration conditions also play a major role in the development of clogging, as two opposing forces are at work. During filtration, an attractive force generated by the suction pump will promote clogging. Parameters that increase the effect of this attractive force are filtration time, filtration flow and TSS concentration. In opposition to this force, a backscattering force tends to limit the deposition of particles on the membrane surface and/or to remove the deposited material. This force is governed by the intensity of aeration, also known as backwashing (Bacchin et al. 1994; Aoustin et al. 2001; Bai and Tien 2005).

There are two main types of clogging: reversible and irreversible. Reversible clogging can be removed in situ by various mechanical methods, whereas irreversible clogging requires chemical cleaning of the membrane.

Plant sizing is based on average net flows of between 15 and 20 $L.h^{-1}.m^{-2}$, with peaks of up to 40 $L.h^{-1}.m^{-2}$.

In addition to the various operating cycles, there is also a cleaning cycle, known as maintenance cleaning, which involves injecting a chlorine or acid solution into the membrane compartment. Pressure maintenance cleaning is triggered automatically. The maximum duration of a maintenance cleaning with bleach is 30 min, and the frequency is once a day.

The maximum duration of an acid maintenance cleaning is 45 min, and the frequency is once a week.

Lastly, the membranes also undergo in-depth cleaning, known as exceptional cleaning. These are carried out manually in a special tank adjacent to the biological tank. This tank is positioned in the axis of the cassette installation, so that these cleaning operations can be carried out by a gantry crane.

The frequency of these exceptional cleanings is around 1 time/year/cassette. The duration of an exceptional cleaning varies from 12 to 24 h per cassette (the soaking time depends on membrane clogging). The operator cleans the aerators, removing the sludge accumulated between the fibers. Washing is carried out with a gentle jet of water. Once the case is washed, it is plunged into the exceptional cleaning tank, previously filled with chlorine solution, and then soaked in the acid solution.

During the backwashing, degassing and relaxation phases, permeate extraction is stopped. This results in a rise in level in the biological tank. As these phases are very short, the increase in level is very small (a few centimeters) and the tank can continue to be fed.

For maintenance cleaning cycles lasting less than 1 h, the water supply to the biological tank is stopped. The water to be treated is then stored in the buffer tank.

Membrane service life is 3–5 years on average, depending on the manufacturer.

3.4.2. *Performance*

3.4.2.1. *Removal of microorganisms*

The ability of MF and UF membranes to remove bacteria and protozoa is well established. UF offers an additional barrier to viruses, whereas MF hardly exceeds 2 log. UF can handle up to 6 logs of bacterial and viral removal (minimum 4 logs). Although MF pore size is larger than virus size, it can remove some viruses when they are stuck to suspended solids and colloids in wastewater.

Beyond its advantages, UF is currently the best approach for removing turbidity from water, a fraction of humic substances and microorganisms, including parasites, and the majority of viruses (Figure 3.22). The final choice of UF solution is often based on these considerations.

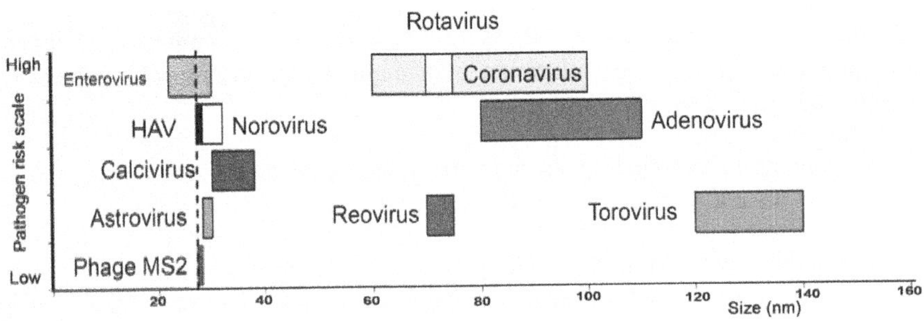

Figure 3.22. *Size of viruses present in water and associated pathogenic risk*

3.4.2.2. *Virus removal*

Virus retention with low-pressure membranes (MF and UF) depends on membrane characteristics, module design and operating mode. Virus removal efficiency can be expressed by an LRV, which is defined as the ability of a filtration system to remove particles (in this case, particles larger than 0.2 μm). LRV is given by:

$$\text{LRV} = \log\left(\frac{Co}{Cs}\right)$$

where Co is the concentration in the feed water and Cs is the concentration in the filtered water.

Viruses are very small (10–400 nm) and should be well retained (at least 4 log) by an integral UF membrane characterized by a MWCO < 150 kDa. Although this parameter is a good indicator, retention cannot be solely predicted on the basis of nominal pore size or MWCO.

Advantages of microfiltration	Advantages of ultrafiltration
– Elimination of microorganisms, including *Cryptosporidium* and *Giardia*. – Turbidity stabilized at less than 0.2 NFU. Removal of dissolved matter (including organic matter) is zero.	– Elimination of microorganisms, including *Cryptosporidium* and *Giardia*. – Elimination of viruses. – Stabilization of turbidity at less than 0.2 NFU. – Reduction in disinfection reagents. – Reduction in trihalomethane (THM) formation. Generally speaking, however (for ultrafiltration with a cut-off of 100 to 150 kDa), there is no reduction in dissolved matter and very little reduction in organic matter (around 10%). A reduction in organic matter can be observed when the latter is made up of very high molecular weight organic matter.

Table 3.3. *Advantages of microfiltration and ultrafiltration*

3.4.2.3. *Process guarantees*

	Microfiltration (100% of the time)	Ultrafiltration (100% of the time)
Turbidity	≤0.2 NTU	≤0.2 NTU
Bacteria	≥5 log removal	≥5 log removal
Cryptosporidium	≥5 log removal	≥5 log removal
Giardia	≥5 log removal	≥5 log removal
Viruses	<2 log	≥4 log removal

Table 3.4. *Examples of process guarantees in MF/UF*

When the membranes are properly operated, and there is no failure or breakage of the membrane fibers, excellent physicochemical and microbiological performance is achieved with a UF membrane, as shown in Tables 3.4 and 3.5.

Parameters	Raw water quality (average) mg.L^{-1}	Treated water quality (average) mg.L^{-1}
	Physicochemical	
COD	448	26.5
BOD5	198	≤5
TSS	224	≤2
TN	42.6	7.8
N-NH$_4$	33.1	0.6
N-NO$_2$	0.1	0.0
N-NO$_3$	0.2	6
TP	9.1	0.3
	Microbiological	
	Raw water (CFU.100 mL^{-1})	**Log removal**
Total coliforms	59×10^6	6.1
Fecal coliforms	17×10^6	6.3
Fecal streptococci	9×10^6	6.3
F-specific RNA bacteriophages	1.5×10^3	3.8

Table 3.5. *Physicochemical and microbiological results obtained at a MBR station (France)*

Municipalities are keen to reuse treated wastewater, as it can be used for almost any purpose apart from direct drinking water.

The benefits of high water quality and low environmental impact have accelerated the development of MBR technology for municipal applications (irrigation, bathing water, urban non-drinking water, etc.).

Effluent quality often far exceeds what is required, and can therefore be seen as a sub-optimal use of scarce resources.

Operating and maintenance costs are higher than for conventional plants. These expenses are linked to membrane systems, and include membrane replacement, energy costs and chemical products. However, the uses to which they are put offset these expenses, as they avoid the need to use drinking water. According to a report analyzing the MBR market, the market for this technology is set to grow in the coming years, given the water restrictions that will be emerging over the next few years.

3.5. Membrane layout in the reuse process

MF and UF membrane layouts downstream of a wastewater treatment plant are diverse and varied. They depend on the uses of the reused water and the microbiological treatment objectives.

1) Application: bathing water protection, irrigation (type B or C) and wetlands.

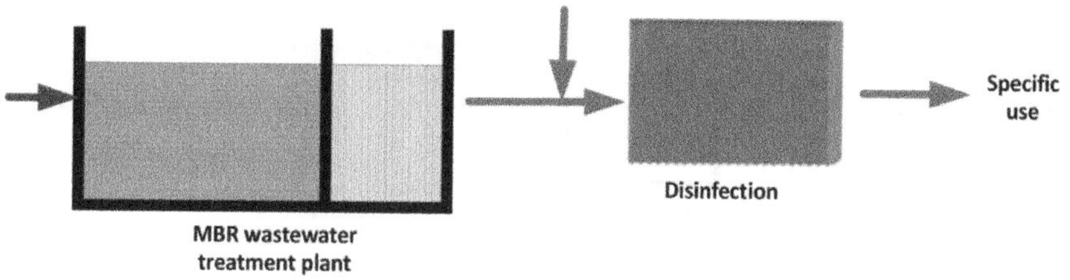

Figure 3.23. *MBR and disinfection*

2) Application: bathing water protection, irrigation (type A, B or C), floor cleaning, urban non-drinking water, wetlands, recreational areas, toilet flushing and specific industries.

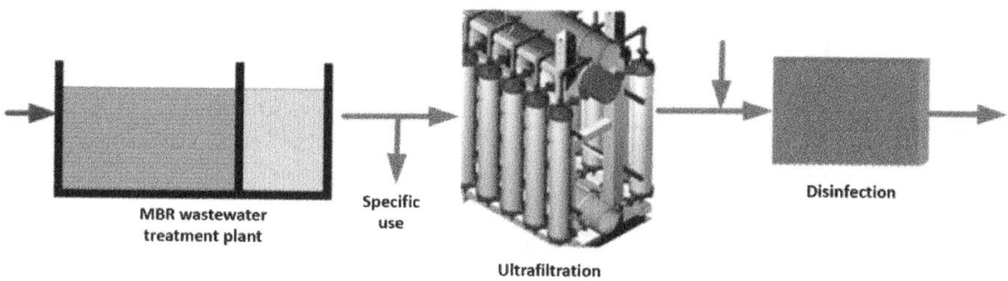

Figure 3.24. *MBR followed by ultrafiltration*

3) Application: bathing water protection, irrigation (type B or C), floor cleaning, urban non-drinking water, wetlands, recreational areas and specific industries.

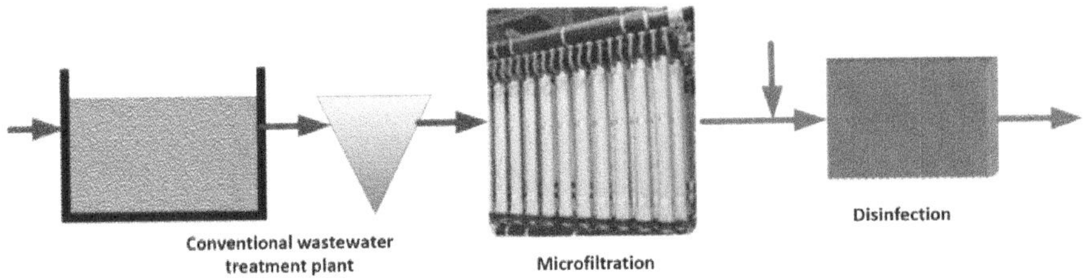

Figure 3.25. *Microfiltration installed downstream of a wastewater treatment plant*

4) Application: bathing water protection, irrigation (type B or C), urban non-drinking water, wetlands, recreational areas and specific industries.

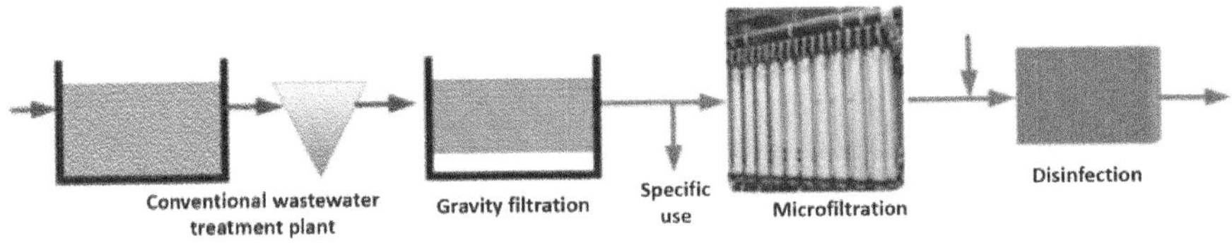

Figure 3.26. *Gravity filtration upstream of microfiltration*

5) Application: bathing water protection, irrigation (type A, B or C), floor cleaning, urban non-drinking water, wetlands, recreational areas, toilet flushing and specific industries.

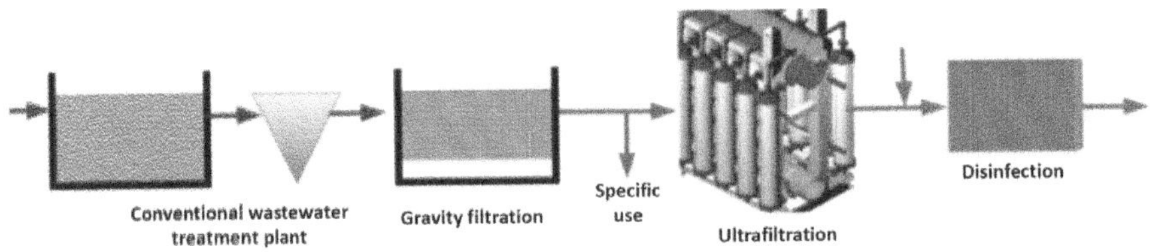

Figure 3.27. *Gravity filtration upstream of ultrafiltration*

6) Application: bathing water protection, irrigation (type A, B or C), floor cleaning, urban non-drinking water, wetlands, recreational areas, toilet flushing, specific industries and groundwater recharge.

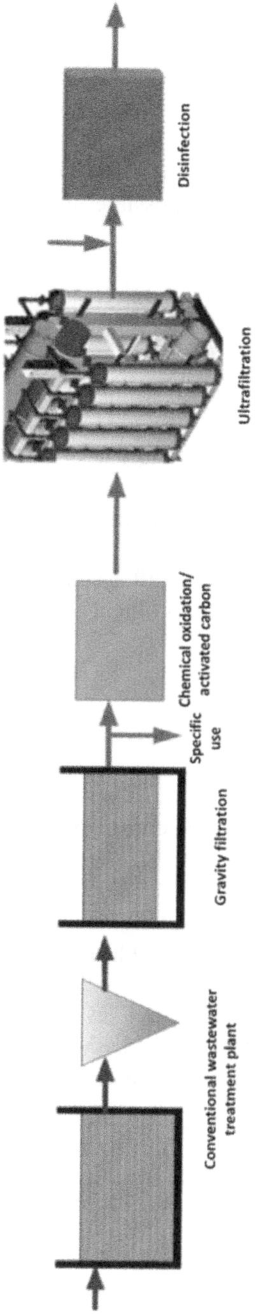

Figure 3.28. *Gravity filtration and advanced treatment upstream of UF*

3.6. Hybrid process: combining chemical processes with MF/UF membranes

AC in powdered (PAC), dispersed or granular (GAC) form is often combined with membranes to reduce organic constituents in the water to be reused.

It contributes to additional removal of organic carbon and micropollutants (see Chapter 2). It also helps to reduce possible membrane fouling and improve membrane filtration performance.

As is already the case in drinking water treatment, the PAC/UF or PAC/MF combination can be used to remove DOC and micropollutants on a one-off basis. This is one of the advantages of this hybrid solution, which can be implemented when the quality of the water to be reused requires it.

UF permeate as the main reuse technology is commonly reused for non-drinking applications, such as agricultural irrigation, landscape irrigation, urban reuse, car washing and as process water for specific industries.

In short, the use of MF or UF alone, after suitable secondary treatment or in a hybrid combination, makes it possible to obtain good quality produced water for a wide range of non-drinking uses.

3.7. Conclusion

MF and UF processes for the reuse of municipal wastewater produce water quality for a variety of uses, meeting the criteria of European directives, the USEPA and the World Health Organization.

Thanks to their performance in removing bacteria, protozoa and viruses (in the case of UF), these processes reduce potential health safety risks.

However, for more specific uses, such as groundwater recharge, raw food irrigation, drinking water production and the food industry, complementary technologies such as NF and reverse osmosis are required. This is the subject of the next chapter.

3.8. References

Abdessemed, D., Nezzal, G., Ben Aim, R. (2000). Coagulation–adsorption–ultrafiltration for wastewater treatment and reuse. *Desalination*, 131, 307–314.

Ahn, K.-H. and Song, K.-G. (1999). Treatment of domestic wastewater using microfiltration for reuse of wastewater. *Desalination*, 126, 7–14.

Altmann, J. and Ripperger, S. (1989). Experimental and theoretical study of the influence of particle interaction forces in crossflow micro- and ultrafiltration. *Sep. Sci. Technol.*, 24, 383–398.

Aoustin, E., Schafer, A.I., Fane, A.G., Waite, T.D. (2001). Ultrafiltration of natural organic matter. *Separation and Purification Technology*, 22(23), 63–78.

Atasoy, E., Murat, S., Baban, A., Tiris, M. (2007). Membrane bioreactor (MBR) treatment of segregated household wastewater for reuse. *Clean Soil Air Water*, 35, 465–472.

Bacchin, P. (1994). Formation et résistance au transfert d'un dépôt de colloïdes sur une membrane d'ultrafiltration. PhD Thesis, Université Paul Sabatier, Toulouse.

Bacchin, P. (2005). Principes de base des technologies à membranes. PhD Thesis, École d'Eté Franco Maghrébine, Monastir.

Bacchin, P., Meireles, M., Aimar, P. (2002). Modelling of filtration: From the polarised layer to deposit formation and compaction. *Desalination*, 145, 139–146.

Bai, R. and Tien, C. (2005). Further work on cake filtration analysis. *Chemical Engineering Science*, 60, 301–313.

Bessiere, Y. (2005). Filtration frontale sur membrane : mise en évidence du volume filtré critique pour l'anticipation et le contrôle du colmatage. PhD Thesis, Université Paul Sabatier - Toulouse III, Toulouse.

Best, G., Singh, M., Mourato, D., Chang, Y.J. (2001). Application of immersed ultrafiltration membranes for organic removal and disinfection by-products reduction. *Water Supply*, 5(6), 221–231.

Bian, R., Watanabe, Y., Tambo, N., Ozawa, G. (1999). Removal of humic substances by UF and NF membranes systems. *Water Science and Technology*, 40(9), 121–129.

Bowen, W.R. and Jenner, F. (1995). Dynamic ultrafiltration model for charged colloidal dispersions: A Wigner-Seitz cell approach. *Chemical Engineering Science*, 50, 1707–1736.

Cabane, B. and Hénon, C. (2003). *Liquides. Solutions, dispersions, émulsions, gels*. Editions Belin, Montparnasse.

Campos, C., Marinas, B.J., Snoeyink, V.L., Baudin, I., Laine, J.M. (2000). PAC membrane filtration process. I: Model development. *Journal of Environmental Engineering*, 126, 97–103.

Carroll, T., King, S., Gray, S.R., Bolto, B.A., Booker, N.A. (2000). The fouling of micro-filtration membranes by NOM after coagulation treatment. *Water Research*, 34(11), 2861–2868.

Chellam, S. and Jacangelo, J.G. (1998). Existence of critical recovery and impact of operational conditions on potable water microfiltration. *Journal of Environmental Engineering*, 124, 1211–1219.

Cho, J., Amy, G., Pellegrino, J. (2000). Membrane filtration of natural organic matter: Factors and mechanisms affecting rejection and flux decline with charged ultrafiltration (UF) membrane. *Journal of Membrane Science*, 164, 89–110.

Choi, K.Y-J. (2003). Critical flux, resistance and removal of contaminants in ultrafiltration of natural organic materials. PhD Thesis, Pennsylvania State University.

Choi, K.Y.-J. and Dempsey, B.A. (2004). In-line coagulation with low-pressure membrane filtration. *Water Research*, 38, 4271–4281.

Crozes, G., Anselme C., Mallevialle J. (1993). Effect of adsorption of organic matter on fouling of ultrafiltration membranes. *Journal of Membrane Science*, 84, 61–77.

Espinasse, B. (2003). Approche théorique et expérimentale de la filtration tangentielle de colloïdes : flux critique et colmatage. PhD Thesis, Université Paul Sabatier, Toulouse.

Espinasse, B., Bacchin, P., Aimar, P. (2002). On an experimental method to measure critical flux in ultrafiltration. *Desalination*, 146, 91–96.

Fabris, R., Lee, E.K., Chow, C.W.K., Chen, V., Drikas, M. (2007). Pre-treatments to reduce fouling of low pressure microfiltration (MF) membranes. *J. Membr. Sci.*, 289, 231–240.

Falsanisi, D., Liberti, L., Notarnicola, M. (2010). Ultrafiltration (UF) pilot plant for municipal wastewater reuse in agriculture: Impact of the operation mode on process performance. *Water*, 2, 872–885.

Fan, L., Harris, J.L., Roddick, F.A., Booker, N.A. (2001). Influence of the characteristics of natural organic matter on the fouling of microfiltration membranes. *Water Research*, 35, 4455–4463.

Farahbakhsh, K. and Smith, D.W. (2004). Removal of coliphages in secondary effluent by microfiltration—Mechanisms of removal and impact of operating parameters. *Water Res.*, 38, 585–592.

Farahbakhsh, K., Svrcek, C., Guest, R.K., Smith, D.W. (2004). A review of the impact of chemical pretreatment on low-pressure water treatment membranes. *J. Environ. Eng. Sci.*, 3, 237–253.

Gómez, M., de la Rua, A., Garralón, G., Plaza, F., Hontoria, E., Gómez, M.A. (2006). Urban wastewater disinfection by filtration technologies. *Desalination*, 190, 16–28.

Gómez, M., Plaza, F., Garralón, G., Pérez, J., Gómez, M.A. (2007). A comparative study of tertiary wastewater treatment by physico-chemical-UV process and microfiltration–ultrafiltration technologies. *Desalination*, 202, 369–376.

Goosen, M.F.A., Sablani, S.S., Al-Hinai, H., Al-Obeidani, S., Al-Belushi, R., Jackson, D. (2004). Fouling of reverse osmosis and ultrafiltration membranes: Critical review. *Separation Science and Technology*, 39(10), 2261–2298.

Goslan, E.H. (2003). Natural organic matter character and reactivity: Assessing seasonal variation in a Moorland water. PhD Thesis, University of Cranfield, Cranfield.

Guigui, C. (2000). Procédé hybride de coagulation/ultrafiltration pour la potabilisation des eaux de surface. PhD Thesis, Université Paul Sabatier, Toulouse.

Guo, W.S., Vigneswaran, S., Ngo, H.H., Chapman, H. (2004). Experimental investigation of adsorption–flocculation–microfiltration hybrid system in wastewater reuse. *J. Membr. Sci.*, 242, 27–35.

Gupta, K. (2021). Fouling of microfiltration membranes during filtration of wastewater effluent for water reuse. PhD Thesis, University of Texas, Texas.

Herath, G., Yamamoto, K., Urase, T. (1999). Removal of viruses by microfiltration membranes at different solution environments. *Water Sci. Technol.*, 40, 331–338.

Hirata, T. and Hashimoto, A. (1998). Experimental assessment of the efficacy of microfiltration and ultrafiltration for Cryptosporidium removal. *Water Science and Technology*, 38, 103–107.

Hong, S. and Elimelech, M. (1997). Chemical and physical aspects of natural organic matter (NOM) fouling of nanofiltration membranes. *Journal of Membrane Science*, 132, 159–181.

Howe, K.J. (2001). Effect of coagulation pretreatment on membrane filtration performance. PhD Thesis, University of Illinois, Urbana-Champaign.

Huang, H., Young, T.A., Schwab, K.J., Jacangelo, J.G. (2012). Mechanisms of virus removal from secondary wastewater effluent by low pressure membrane filtration. *J. Membr. Sci.*, 409(410), 1–8.

Jacangelo, J.G., Rhodes Trussell, R., Watson, M. (1997). Role of membrane technology in drinking water treatment in the United States. *Desalination*, 113, 119–127.

Jucker, C. and Clark, M.M. (1994). Adsorption of aquatic humic substances on hydrophobic ultrafiltration membranes. *Journal of Membrane Science*, 97, 37–52.

Judd, S.J. (2010). *The MBR Book: Principles and Applications of Membrane Bioreactors for Water and Wastewater Treatment*. Elsevier, Amsterdam.

Judd, S.J. and Hillis, P. (2001). Optimisation of combined coagulation and microfiltration for water treatment. *Water Research*, 35, 2895–2904.

Jung, C.W., Son, H.J., Kang, L.S. (2006). Effects of membrane material and pretreatment coagulation on membrane fouling: Fouling mechanism and NOM removal. *Desalination*, 197, 154–164.

Kabsch-Korbutowicz, M. (2006a). Removal of natural organic matter from water by in-line coagulation/ultrafiltration process. *Desalination*, 200, 421–423.

Kabsch-Korbutowicz, M. (2006b). Impact of pre-coagulation on ultrafiltration process performance. *Desalination*, 194, 232–238.

Kecili, K. (2006). Étude et caractérisation de membranes synthétiques organiques immergées de microfiltration (MF) et d'ultrafiltration (UF) employées pour la production d'eau destinée à la consommation humaine : intensification des opérations de déconditionnement et de nettoyage. PhD Thesis, Université Pierre et Marie Curie, Paris.

Kennedy, M., Kim, S.M., Mutenyo, I., Broens, L., Schippers, J. (1998). Intermittent crossflushing of hollow fiber ultrafiltration systems. *Desalination*, 118, 175–187.

Kim, J.-S., Akeparathumchai, S., Wickramasinghe, S.R. (2001). Flocculation to enhance microfiltration. *Journal of Membrane Science*, 182, 161–172.

Kim, K.-Y., Kim, H.-S., Kim, J., Nam, J.-W., Kim, J.-M., Son, S. (2009). A hybrid microfiltration–granular activated carbon system for water purification and wastewater reclamation/reuse. *Desalination*, 243, 132–144.

Kulovaara, M., Metsamuuronen, S., Nystrom, M. (1999). Effects of aquatic humic substances on a hydrophobic ultrafiltration membrane. *Chemosphere*, 38, 3485–3496.

Kwon, B., Park, N., Cho, J. (2005). Effect of algae on fouling and efficiency of UF membranes. *Desalination*, 179, 203–214.

Laabs, C., Amy, G., Jekel, M. (2004). Organic colloids and their influence on low-pressure membrane filtration. *Water Science and Technology*, 50, 311–316.

Laine, J.-M., Vial, D., Moulart, P. (2000). Status after 10 years of operation–overview of UF technology today. *Desalination*, 131, 17–25.

Langlet, J., Ogorzaly, L., Schrotter, J.-C., Machinal, C., Gaboriaud, F., Duval, J.F.L., Gantzer, C. (2009). Efficiency of MS2 phage and Q phage removal by membrane filtration in water treatment: Applicability of real-time RT-PCR method. *J. Membr. Sci.*, 326, 111–116.

Lee, N., Amy, G., Habarou, H., Schrotter, J.C. (2003). Identification and control of fouling of low pressure (MF and UF membranes) by drinking-water natural organic matter. *Water Supply*, 3(5), 217–222.

Lee, N., Amy, G., Croue, J.P., Buisson, H. (2004). Identification and understanding of fouling in low pressure membrane (MF/UF) filtration by natural organic matter (NOM). *Water Research*, 38, 4511–4523.

Lee, C.W., Bae, S.D., Han, S.W., Kang, L.S. (2007). Application of ultrafiltration hybrid membrane processes for reuse of secondary effluent. *Desalination*, 202, 239–246.

Liang, H. (2007). Study on algae monitoring/prediction and its control mechanisms. PhD Thesis, Harbin Institute of Technology.

Liang, H., Gong, W.J., Chen, Z.L., Tian, J.Y., Qi, L., Li, G.B. (2009). Effect of chemical preoxidation coupled with in-line coagulation as a pretreatment to ultra-filtration for algae fouling control. *Desalination and Water Treatment*, 9, 241–245.

Lin, C.F., Huang, Y.J., Hao, O.J. (1999). Ultrafiltration processes for removing humic substances: Effect of molecular weight fractions and PAC treatment. *Water Research*, 33, 1252–1264.

Lin, C.F., Lin, T.Y., Hao, O.J. (2000). Effects of humic substance characteristics on UF performance. *Water Research*, 34, 1097–1106.

Liu, P., Hill, V.R., Hahn, D., Johnson, T.B., Pan, Y., Jothikumar, N., Moe, C.L. (2012). Hollow-fiber ultrafiltration for simultaneous recovery of viruses, bacteria and parasites from reclaimed water. *J. Microbiol. Methods*, 88, 155–161.

Lv, W., Zheng, X., Yang, M., Zhang, Y., Liu, Y., Liu, J. (2006). Virus removal performance and mechanism of a submerged membrane bioreactor. *Process Biochem.*, 41, 299–304.

Maartens, A., Swart, P., Jacobs, E.P. (2000). Membrane pretreatment: A method for reducing fouling by natural organic matter. *Journal of Colloid and Interface Science*, 221, 137–142.

Madaeni, S.S., Fane, A.G., Grohmann, G.S. (1995). Virus removal from water and wastewater using membranes. *J. Membr. Sci.*, 102, 65–75.

Martin-Mousset, B., Croue, J.P., Lefebvre, E., Legube, B. (1997). Distribution and characterization of dissolved organic matter of surface waters. *Water Research*, 31, 541–553.

Matsui, Y., Yuasa, A., Ariga, K. (2001a). Removal of a synthetic organic chemical by PAC-UF systems. I: Theory and modeling. *Water Research*, 35, 455–463.

Matsui, Y., Colas, F., Yuasa, A. (2001b). Removal of a synthetic organic chemical by PAC-UF systems. II: Model application. *Water Research*, 35, 464–470.

Miljatovic, I., Matosic, M., Hajduk Cerneha, B., Bratilic, D. (2004). Removal of natural organic mater by ultrafiltration and nanofiltration for drinking water production. *Desalination*, 169, 223–230.

Mo, L. and Huang, X. (2003). Fouling characteristics and cleaning strategies in coagulation-microfiltration combination process for water purification. *Desalination*, 159, 1–9.

Mulder, M. (ed.) (1996). Polarisation phenomena and membrane fouling. In *Basic Principles of Membrane Technology*. Kluwer Academic Publishers, Dordrecht.

Muthukumaran, S., Nguyen, D.A., Baskaran, K. (2011). Performance evaluation of different ultrafiltration membranes for the reclamation and reuse of secondary effluent. *Desalination*, 279, 383–389.

Otaki, M., Yano, K., Ohgaki, S. (1998). Virus removal in a membrane separation process. *Water Sci. Technol.*, 37, 107–116.

Ottoson, J., Hansen, A., Björlenius, B., Norder, H., Stenström, T.A. (2006). Removal of viruses, parasitic protozoa and microbial indicators in conventional and membrane processes in a wastewater pilot plant. *Water Res.*, 40, 1449–1457.

Park, P.K., Lee, C.H., Choi, S.J., Choo, K.H., Kim, S.H., Yoon, C.H. (2002). Effect of the removal of DOMs on the performance of a coagulation – UF membrane system for drinking water production. *Desalination*, 145, 237–245.

Purnell, S., Ebdon, J., Buck, A., Tupper, M., Taylor, H. (2016). Removal of phages and viral pathogens in a full-scale MBR: Implications for wastewater reuse and potable water. *Water Res.*, 100, 20–27.

Rickman, M., Pellegrino, J., Davis, R. (2012). Fouling phenomena during membrane filtration of microalgae. *Journal of Membrane Science*, 423(424), 33–42.

Seo, G.T., Suzuki, Y., Ohgaki, S. (1996). Biological powdered activated carbon (BPAC) microfiltration for wastewater reclamation and reuse. *Desalination*, 106, 39–45.

Serra, C. (1996). Ultrafiltration frontale par fibres creuses : dimensionnement des modules et simulation du procédé en clarification des eaux. PhD Thesis, Université Paul Sabatier, Toulouse.

Shanmuganathan, S., Johir, M.A.H., Nguyen, T.V., Kandasamy, J., Vigneswaran, S.L. (2015). Experimental evaluation of microfiltration–granular activated carbon (MF–GAC)/nano filter hybrid system in high quality water reuse. *J. Membr. Sci.*, 476, 1–9.

Wetterau, G.E., Clark, M.M., Anselme, C. (1996). A dynamic model for predicting fouling effects during the ultrafiltration of a groundwater. *Journal of Membrane Science*, 109, 185–204.

Yuan, W. and Zydney, A.L. (2000). Humic acid fouling during ultrafiltration. *Environmental Science and Technology*, 34, 5043–5050.

Zhang, X., Hu, Q., Sommerfeld, M., Puruhito, E., Chen, Y. (2010). Harvesting algal biomass for biofuels using ultrafiltration membranes, *Bioresource Technology*, 101, 5297–5304.

Zheng, X., Lü, W., Yang, M., Liu, J. (2005). Evaluation of virus removal in MBR using coliphages T4. *Chin. Sci. Bull.*, 50, 862–867.

4

Reverse Osmosis

While MF and UF membrane technologies are based on processes that use microporous membranes that operate under pressure to remove particles (including turbidity and micro-organisms) by a size exclusion mechanism, reverse osmosis (RO) membrane technology is based on the use of semi-permeable membranes, which operate under pressure to primarily target the removal of dissolved contaminants (organics and ions) by a controlled diffusion separation process.

Thus, when a high pressure is applied to the feed side of the membrane, above the system's natural osmotic gradient, water is forced through the molecular structure of the membrane surface. Most of the contaminants are then rejected. As the mass transfer rate of these contaminants is slower than that of water, the result is a permeate that is largely free of these compounds.

The amount of energy (hydraulic pressure) required to drive the feed water through the membrane depends on the membrane material and thickness, as well as the osmotic pressure of the feed water. Osmotic pressure is the pressure created by the natural process of water flowing from a dilute solution (in other words, one with a low concentration of compounds) through a semi-permeable membrane to a more concentrated solution (that is to say, one with a higher concentration of compounds). Thus, energy in the form of hydraulic pressure is required to overcome both the physical resistance of the membrane and the osmotic pressure. As the applied pressure works against the osmotic pressure, the treatment process is called RO. Under given operating conditions, the feed pressure primarily depends on the total dissolved solids (TDS) concentration and the temperature of the feed water. Thus, osmotic pressure (π) is the pressure which, if applied to the most concentrated solution, would prevent water transport across the membrane.

4.1. Membranes

The most widely used suppliers of RO membranes today are Toray (Japan), Dow (USA), Hydranautics (USA) and Koch (USA). The membranes currently on the market are exclusively spiral-wound and cannot be backwashed.

4.1.1. *Materials*

The two types of membrane currently in use are asymmetrical homogeneous membranes and composite membranes. The main materials used to manufacture RO membranes are based on various polyamides (notably aromatic polyamides).

RO membranes are organic and composite, that is to say, composed of several distinct layers:

– a porous polyester textile backing (around 120 µm thick). This support gives the membrane its mechanical strength;

– a layer of microporous polysulfone (PS; pore diameter around 15 mm), around 40 µm thick. This layer provides an extremely smooth surface for adhesion of the third layer;

– a very thin skin (ultra-fine barrier), around 0.2–2 µm thick, which gives the membrane its specific rejection properties for the various compounds. The composition of this layer varies from one membrane to another (polypiperazine-amide, PS, etc.).

The water to be filtered is in contact with the last layer and flows from the thinnest to the loosest layer. As a result of this particular composite structure, it is not possible to reverse the direction of water flow (from permeate to concentrate): this would risk detaching the different layers of the membrane. Consequently, a positive pressure difference must always be maintained between concentrate and permeate. The surface, or skin, gives the membrane its rejection characteristics, and the porous substrate supports this skin to withstand pressure differences of over 70 bar in some seawater RO applications. In wastewater, where salt concentrations are much lower, the pressure is more likely to be between 12 and 18 bar. Support materials commonly used for thin-film composite (TFC) membranes include PS, polyethersulfone (PES), polyvinylidene fluoride (PVDF), polyacrylonitrile (PAN) and polyetheretherketone (PEEK).

Polymer membranes account for the largest number of membranes installed in drinking water and wastewater treatment plants. Depending on the manufacturer, different polymers are used to ensure optimum operating conditions.

Tubular designs provide excellent flow distribution and high sweep speed, which minimizes concentration bias and pre-treatment conditions, but also means high pumping costs. This is why tubular modules are usually used for specialized, low-flow operations.

The industrial modules have a diameter of 203.2 mm and a length of 1,016 mm, and feature a hydraulic efficiency (conversion rate) of around 15%.

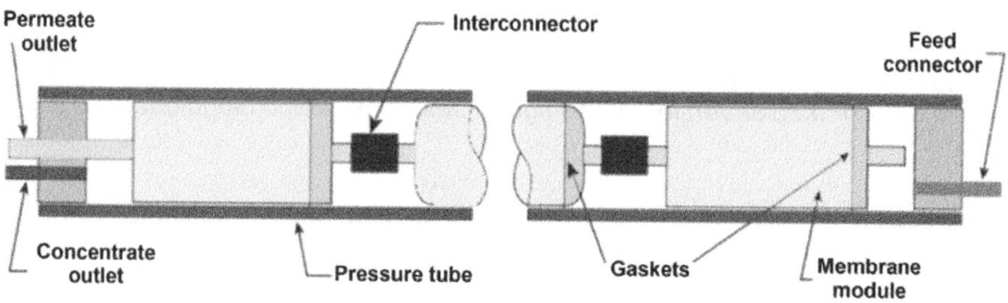

Figure 4.1. *Pressure tube assembly*

Figure 4.2. *Interconnectors in a high-pressure membrane. For a color version of this figure, see www.iste.co.uk/gaid/reuse2.zip*

Each pressure tube consists of six to seven membranes linked together by a permeate connector, which connects the permeate collectors to one another (Figure 4.1). The tubes are then closed by components known as bottom plates. These are connected to the high-pressure collector (feed and concentrate) by Victaulic-type connectors. The pressure tubes are made up of stainless steel with side port connections and are fed from the side (Figure 4.3).

Figure 4.3. *Various pressure tube components (baseplate a), Victaulic b) and feed c)). For a color version of this figure, see www.iste.co.uk/gaid/reuse2.zip*

Spiral-wound membranes are standardized and interchangeable from one supplier to another.

4.2. Principles of operation and separation

In membrane processes, a membrane separates two phases by acting as a selective barrier, regulating the transport of substances between the two compartments. The barrier structure of membranes can be classified according to their porosity. The membrane surface can be a dense selective skin, allowing only diffusive transport, or a porous skin, allowing permeate flow. For porous membranes, transport speed and selectivity are mainly influenced by viscous flow and steric exclusion. However, solute interactions with the porous membrane surface can alter membrane performance.

The main advantages of membrane technology over other unit operations are linked to this unique separation principle, namely, the membrane's transport selectivity. Membrane separations require no additives, can be carried out isothermally at low temperatures, and can be combined with other low-energy thermal separation processes.

The RO membrane process uses the pressure difference between the feed and permeate sides as the driving force to transport water across the membrane. Dissolved compounds are partially or totally retained, depending on properties such as size exclusion, charge repulsion or dielectric exclusion. Separation may also be due to chemical interactions such as hydrophobic interactions or hydrogen bonds.

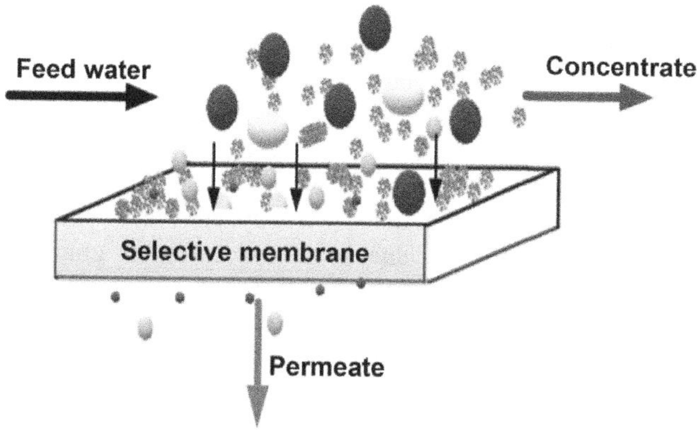

Figure 4.4. *Simplified conceptual diagram of membrane separation*

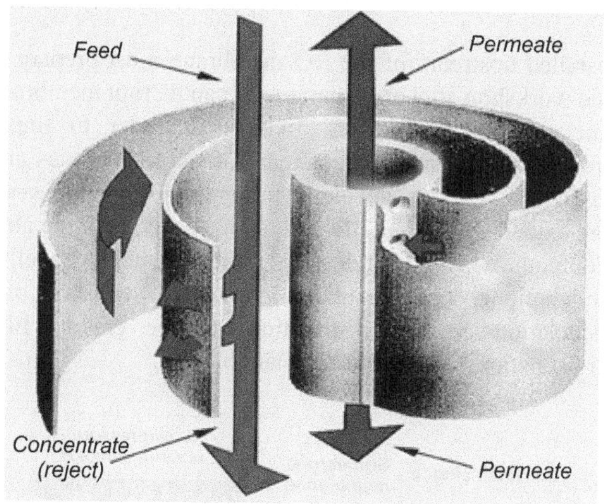

Figure 4.5. *Separation into two streams (permeate and concentrate (reject))*

The mechanisms can be explained by size/steric exclusion, hydrophobic adsorption and separation, and electrostatic repulsion, given the identification of the compound's physicochemical properties and membrane characteristics in transport, adsorption and removal.

Thus, the impact of molecular size concerns most organic compounds. All those with a molecular weight (MW) above the membrane cut-off (MWCO) are rejected by the membrane. Those with a MW below a membrane's MWCO can penetrate the membrane if they are not held back by physicochemical interactions.

Interactions due to Van der Waals forces are mainly responsible for the retention of uncharged molecules. They are transported by convection due to a pressure difference, and also by diffusion due to a concentration gradient across the membrane.

For charged compounds, size exclusion and electrostatic interactions are responsible for separation.

4.3. Wastewater treatment with reverse osmosis membranes

The treatment process depends on the quality of the raw water and is designed on the basis of the parameters considered in section 4.4.

Processes installed upstream of the RO membrane must prepare the feed water for the membrane workshop so that no parameter can disrupt membrane operation or fail to meet the required guarantees. Indeed, in order to improve filtration performance, pre-treatment processes upstream of the membranes are necessary to reduce the potential for clogging development and also improve system operation. Pre-treatment can extend membrane life and reduce costs, with the ability to remove more organic compounds, soluble microbial products and extracellular polymeric substances. Conventional pre-treatment processes consist of clarification (coagulation-flocculation-settling or flotation) and/or gravity filtration and/or activated carbon ozonation-filtration at the plant outlet.

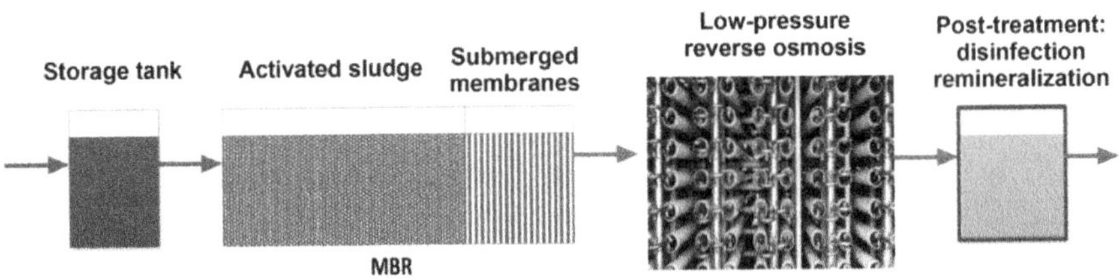

Figure 4.6. *Pre-treatment with a MBR upstream of the RO membrane train*

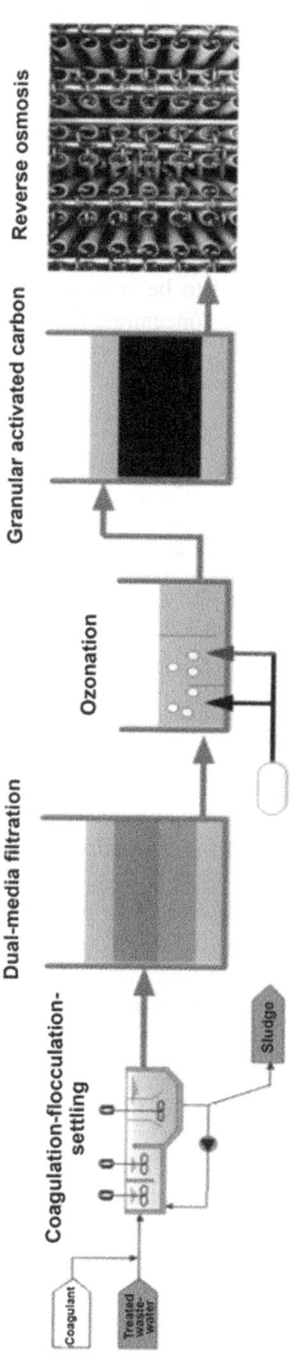

Figure 4.7. *Pre-treatment with conventional treatment upstream of the RO membrane train*

Pre-treatments including membranes are more efficient than conventional pre-treatments and produce better water quality. An MF or UF membrane placed downstream of a treatment plant is less risky than an MBR solution, because in the event of failure of the MBR membranes, the raw wastewater must be discharged into the river until the plant malfunctions or the damage is repaired. In contrast, with a downstream membrane, any momentary failure of this system means that treated wastewater can be temporarily discharged, causing less damage to the environment and surrounding area.

In all cases, three water flows need to be considered: the raw water flow, the produced water flow (permeate) and the concentrate flow.

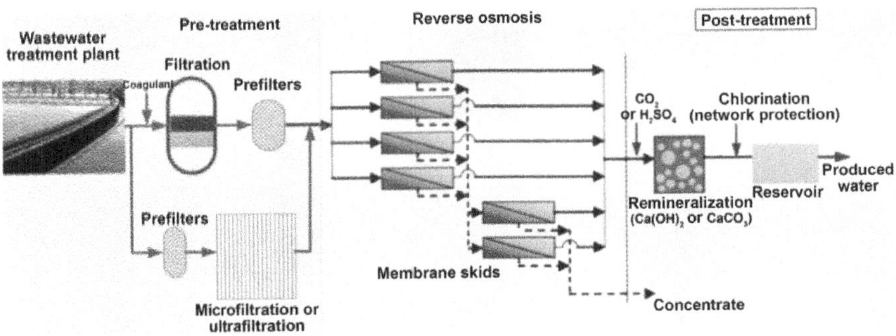

Figure 4.8. *Pre-treatment with MF or UF membrane upstream of the RO membrane train*

4.3.1. *Parameters to consider when designing a treatment system*

The separation characteristics of low-pressure reverse osmosis (LPRO) include size exclusion, charge exclusion, diffusion mechanisms and electrostatic interactions. As a result, the feedwater requirements of RO membranes are relatively higher than those of MF and UF membranes, because small particles such as natural organic matter and others are likely to cause early clogging. In wastewater, biofouling, represented by the irreversible adhesion to a membrane caused by microorganisms and extracellular polymers, is considered to be the RO operation's main obstacle over the medium to long term.

The parameters affecting the operation of the RO plant are turbidity, TDS, SDI (silt density index), pH, as well as temperature, which affects viscosity and requires all flow and permeability values to be corrected to a standard temperature of 20°C.

4.3.1.1. *Particulates and TSS*

RO acts on dissolved substances and is not a suitable process for retaining particulate pollution. Particulate pollution must therefore be treated as efficiently as possible upstream of the membranes in order to minimize clogging.

Turbidity can give an initial indication, but it is limited in assessing the quality of pre-treatment with regard to membranes.

4.3.1.2. *Particle count*

Most meters count particles as small as 1.5 μm, with several channels providing a "particle size distribution". Some meters go down to 0.5 μm. However, in practice, the variation in values obtained has often led operators to minimize the use of this equipment.

4.3.1.3. *SDI or MFI*

4.3.1.3.1. Clogging indicators

Two indices are commonly used to estimate the clogging capacity of membrane feedwater: the SDI and modified fouling index (MFI). The SDI is more regularly used, as it is easier to implement. Both indices are based on the principle of Darcy's law applied to cake filtration, and are determined by tracking the loss of flux through a flat membrane with a porosity of 0.45 μm (filtering surface area of 1.73×10^{-4} m^2) at a pressure of 2.07 bar. The duration of the experiment varies from a few minutes to an hour, depending on the nature of the water. High indices are generally associated with a rapid decrease in membrane performance.

The difference between MFI and SDI is that the former is directly related to the concentration of clogging particles and is informative about cake-type clogging, whereas the latter is more global and is not solely related to cake-type clogging.

4.3.1.3.2. Silt density index

The clogging index most often used by the water industry and membrane manufacturers is the SDI.

The SDI is therefore representative of a particle density that qualifies a water's clogging properties. It is calculated from the time required to filter a given volume of water through a flat membrane with a pore diameter of 0.45 μm (microfiltration [MF]) and a filtering surface area of 1.73×10^{-4} m^2 at a constant transmembrane pressure of 2.07 bar. The membrane flow is frontal. The measuring principle consists of recording the amount of time required (t_1) to filter 500 mL of water through a new test membrane, then, after a filtration time T (between 1 and 60 min), to again record the time required (t_2) to filter another volume of 500 mL.

The SDI index is expressed in %/minute. It is calculated by:

$$SDItf = \frac{100}{tf}\left(1 - \frac{t1}{t2}\right)$$

where:

– t_1 is the time required to collect the first 500 mL of sample (min);

– t_2 is the time required to collect the last 500 mL of sample (min);

– t_f is the total filtration time (min).

This index only provides an indication of particle clogging and a static measure of filtration resistance. The more clogging the water, the shorter the overall filtration time, as the membrane is quickly clogged.

The choice of test duration depends on the factor $\left(1-\frac{ti}{t_f}\right)$:

– by default and provided that $\left(1-\frac{ti}{t_f}\right) < 0.75$, the duration is 15 min;

– if $\left(1-\frac{ti}{t_f}\right) > 0.75$, then test duration is between 1 and 10 min.

The ranges of validity of the measurements are therefore as follows:

– SDI_{15}: "0" to 6.6;

– SDI_{10}: "0" to 10;

– SDI_5: "0" to 20;

– SDI_3: "0" to 33.3.

The 15-min reference time for t_f is often the benchmark for pre-treated water.

For example, tests on two types of water gave the results shown in Table 4.1.

Type 1 water, with an $SDI_{15} < 3$, is wastewater treated with MF membranes. Type 2 water, with an SDI > 5, is wastewater treated with dual-media filtration.

Time (min)	Volume of water filtered (mL)	Type 1 water Collection time (s)	Type 2 water Collection time (s)	SDI
t_i: 0	500	9	42	–
t_{f1}: 15	500	14	–	2.4
t_{f2}: 5	500	–	75	>5

Table 4.1. *Examples of SDI results for two types of water*

The advantage of SDI is that it is relatively easy to determine. However, it has one major drawback: this test does not take into account the different phases of membrane clogging. First, pores are considered to be blocked, then a cake is formed. Depending on the water being tested, SDI measurements can be taken partially or totally before cake formation.

4.3.1.3.3. Modified fouling index

Compared with SDI, MFI physically represents the resistance of one unit of deposit mass per unit area of filter membrane. Membrane clogging in frontal filtration, where the pressure gradient is constant, takes place in three stages: blocking of pores, formation of an incompressible cake and then formation of a compressible cake. The denser the deposit formed, the more difficult it is for the fluid to be filtered. The MFI is therefore high.

When an incompressible cake is formed, a linear relationship exists between t/V and V, as shown in Figure 4.9. The MFI indicator specifically corresponds to clogging, due to cake formation on the membrane surface, and is represented by the slope of the linear part of the curve t/V = f(V).

$$\mathrm{MFI}0.45 = \frac{\mu 20}{\mu T}\left(\frac{\Delta P}{210}\right)\tan\alpha$$

where V, t, ΔP and μ represent filtrate volume, filtration time, applied transmembrane pressure and dynamic water viscosity, respectively. The MFI is calculated at a temperature of 20°C and a pressure of 210 kPa.

In practice, MFI determination is similar to that of SDI. The same experimental set-up is used. The tests consist of a series of time measurements and cumulative permeate volumes, enabling t/V to be plotted as a function of V.

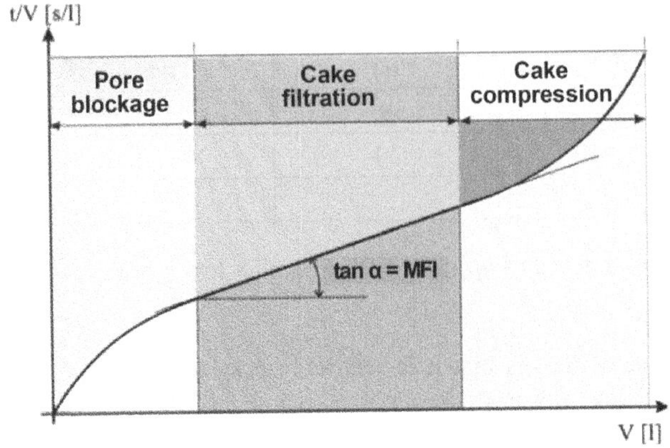

Figure 4.9. *Experimental determination of the MFI*

4.3.1.4. *Salts and metals*

As RO is a salt concentration technique, salts tend to precipitate when their solubility product (Ksp) is exceeded. The higher the conversion rate, the higher the salt concentration. In this case, there is a high risk of exceeding the solubility products of the salts. All the more so as the concentration of salts and metals retained by the membranes increases in the concentrate from top to bottom, which can lead to precipitation and cause clogging.

Certain salts that are present in groundwater or brackish water can limit the plant's conversion rate, particularly in the case of high sulfate concentrations. These include calcium fluoride (CaF_2), barium sulfate ($BaSO_4$), strontium sulfate ($SrSO_4$), calcium sulfate ($CaSO_4$), silica (SiO_2) and calcium carbonate ($CaCO_3$).

Precipitation inhibitors, also known as sequestrants, are generally used to keep salts in solution above their theoretical saturation limit.

With $CaCO_3$, the action of the sequestering agent is not enough: the pH of the feed water also needs to be adjusted.

Sizing an RO unit requires knowledge of the ionic balance of the water to be treated. When a risk of salt precipitation is identified, water conditioning must be provided in the membrane feed water by injecting a dispersant (also known as a sequestrant).

4.3.1.4.1. Clogging of reverse osmosis membranes

One of the most recurrent problems in membrane plants is the progressive degradation of the quality of the permeate produced. The reduction in flow rate can be attributed to increased reversible (concentration polarization) or irreversible (clogging) resistance of the membrane to permeate flow.

The number of studies on the critical, sustainable and threshold flux of different types of wastewater filtered by certain membranes has recently increased, as these flux concepts are strongly linked to the control of low fouling rates on membranes. For polymer membranes, material selection depends on pore size distribution, wetting sensitivity, porosity, mechanical strength, cost, polymer flexibility, fouling resistance, stability, durability and chemical resistance.

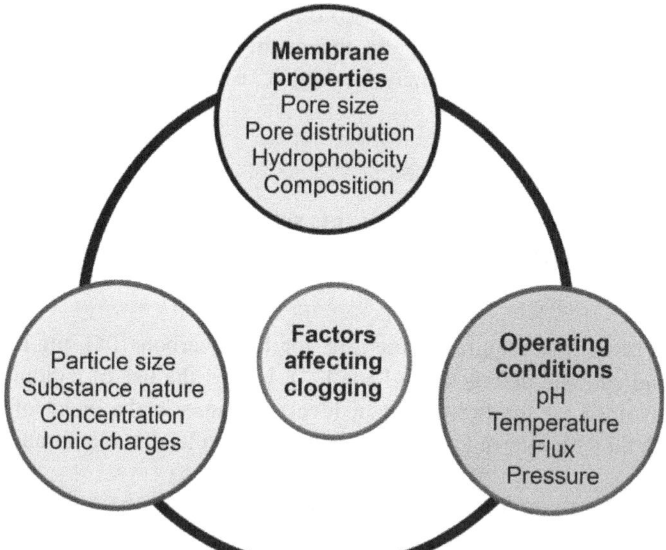

Figure 4.10. *Factors affecting membrane clogging*

Scaling of poorly soluble salts, biological clogging and irreversible clogging are the main limitations of all membrane operations, slowing down their use and further development in both desalination and wastewater treatment.

A reduction in permeate flux is observed when precipitation and/or adsorption mechanisms take place on the membrane surface or in the pores. Calcium sulfate and carbonate deposits are often the main cause of fouling, in addition to complex organic compounds containing acidic functional groups such as carboxyls and alkyl halides, which gradually form a dense layer.

Bacteria can, in some cases, adhere to the membrane and grow, producing a viscous gel consisting of exopolysaccharides that create a "biofilm" that is difficult to remove. The main parameters influencing such developments are:

– the concentration of bacteria or algae in the feed water;

– water temperature;

– the operating mode of the installation: prolonged shutdowns promote proliferation. Appropriate operating procedures help to limit risks (maximum downtime, rinsing, conditioning and operation).

In general, for a wastewater application, the frequency of chemical cleaning is twice a year. However, depending on the quality of the water entering the RO system, the cleaning frequency may be higher, even reaching one cleaning per quarter.

4.3.1.4.2. Undesirable substances

Manufacturers consider a number of substances to be incompatible with membranes. These include oils, greases, non-ionic and cationic detergents, and hydrocarbons.

In most cases, the required limits are for hydrocarbons (<1 $mg.L^{-1}$) and for non-ionic and cationic detergents (<1 $mg.L^{-1}$). Lastly, the use of cationic polymers should be avoided (at the pre-treatment level), because in the event of leakage, a reaction with the sequestering agent is possible and can lead to membrane clogging.

4.3.1.4.3. Limit values for compounds entering reverse osmosis membranes for treated wastewater

Most of the compounds responsible for clogging, as described above, can be removed by chemical cleaning, with the exception of calcium sulfate, barium sulfate and strontium sulfate. For these salts, special sequestrants must be added, but it may not be sufficient.

Table 4.2 gives a rough definition of the different cases of a high-pressure membrane feed. The values should be interpreted as average values. Indeed, a short overrun may require cleaning, with no significant impact on either membrane performance or annual operating costs.

Parameters	Reference limit values	Imperative limit values
Turbidity (NTU)	(<0.2)	<0.5
Particles > 1.5 *µ*	<200	<2,000
SDI	Approximately 3	<5
Algae	Absent	Absent
Al (µg.L^{-1})	<20 (detection limit)	<50
Fe (µg.L^{-1})	<20 (detection limit)	<50
Mn (µg.L^{-1})	<20 (detection limit)	<50

Table 4.2. *Limit values for some acceptable elements and substances in high-pressure membrane feed systems*

4.3.2. *Sizing parameters*

The maximum permeate flux that can be produced depends on the potential for membrane clogging and scaling. A membrane system for wastewater is usually laid out in three stages, with a hydraulic efficiency of between 80 and 85%.

In all cases, the concentrate must be considered to determine its outcome. The concentrations of salts and other contaminants in the concentrate are calculated as a function of the conversion rate. The evaluation of investment and operating costs must take into account the physicochemical quality of the raw water, the costs of reagents (including pre-treatment, membrane station and post-treatment), energy costs, maintenance and membrane renewal costs.

4.3.2.1. *Temperature*

Temperature is the parameter with the greatest influence on performance, both in terms of water quality and inlet pressure. In some countries, wastewater temperatures vary considerably over the course of the year, so sizing must take into account the minimum temperature, which involves calculating maximum pressure, maximum membrane rejection and equilibrium conditions (pH and soda consumption).

4.3.2.2. *Implementation configuration*

Given that RO and nanofiltration are pressure-driven processes, the required membrane feed pressure is a critical hydraulic design parameter. Membrane feed pressure directly influences membrane flux and permeate water quality. The following five parameters influence the efficiency of the RO/NF system:

– osmotic pressure;

– pressure loss through the membrane;

– permeate backpressure;

– pressure loss along the feed channel of multiple membrane elements configured in series within a pressure vessel;

– friction losses in the feed piping between membrane feed stages.

The choice of flux must be weighed up against the performance of the membrane system and economic considerations. A compromise must be found between aggressive sizing, which leads to excessive clogging, and investment costs, by considering a flux that is too low.

The main sizing parameters are as follows:

– permeate flux (Jw, $L.m^{-2}.h^{-1}$)) is the ratio of permeate flow rate (Q) to total membrane surface area (St). It is also proportional to the pressure gradient across the membrane (transmembrane pressure) and to membrane permeability (Lp in $L.m^{-2}.h^{-1}.bar^{-1}$):

$$Jw = \frac{Qp}{St} = \mathrm{Lp}\ (\Delta Pt - \Delta\Pi\ ) = \mathrm{Lp}\ (\Delta P\mathrm{effective})$$

with:

$$\Delta Pt = [\frac{Pin+Pc}{2} - Pp\ ]$$

or even:

$$\Delta P\mathrm{effective} == \left[\frac{Pin + Pc}{2} - Pp\right] - [\frac{\Pi in + \Pi c}{2} pf - \Pi p]$$

S (m^2) is the membrane surface area, Qp ($L.h^{-1}$) is the flow rate produced by the membrane and ΔΠ is the osmotic pressure difference on either side of the membrane, which ranges from 0.1 bar to 1 or 2 bar, depending on the nature of the solution used (water more or less concentrated in salts). ΔΠ (bar) includes the osmotic pressures of the feed water, the permeate and the concentrate.

Pin, Pc and Pp (bar) are inlet pressure, concentrate pressure and permeate pressure, respectively. A temperature coefficient is included to establish the correction with respect to a reference temperature. pf is the polarization factor. Attack pressure (Pin, bars) gradually increases when there is a decrease in flux due to a drop in temperature or clogging.

A compromise must therefore be found, depending on the quality of the feed water, between the conversion rate and the clogging. The latter also depends on the permeate flow. The permeability *Lp* depends on the temperature and the degree of clogging of the membrane. The higher it is, the better the hydraulic performance of the membrane.

The flow rate produced (Qp, $m^3.day^{-1}$) determines the number of modules to be implemented.

In wastewater, applied fluxes are around 18–20 lmh. A two-stage configuration provides a conversion rate of 75–80%, while a three-stage configuration provides at least 85% conversion rate.

Conversion by element (Y_e):

$$Qp / Qin = 1 - (1 - Y_e)^n$$

where n is the number of pressure tubes in the system. It therefore follows that for a stage where each pressure tube comprises m elements, the conversion (Y_{stage}) is given by:

$$Y_{stage} = 1 - (1 - Y_e)^m$$

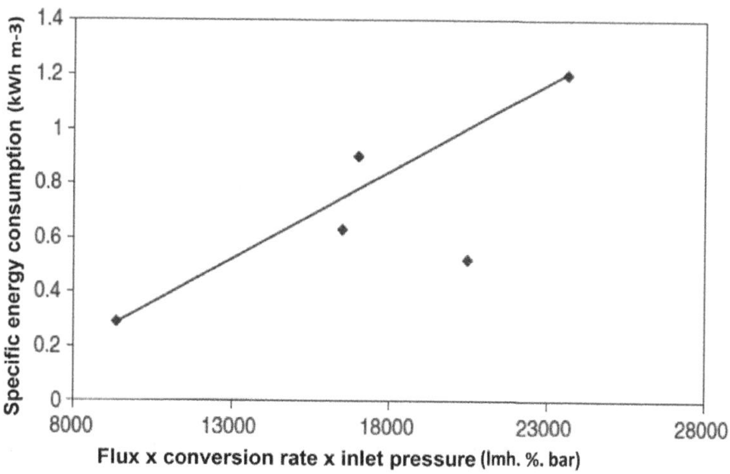

Figure 4.11. *Specific energy consumption as a function of flux/conversion rate/pressure*

The number of elements per pressure tube is six or seven at all sites. For an estimated conversion of 50% for the first stage, as would be normal, the average conversion per element is 9.4–10.9%. This is within the overall per-element conversion range of 8.6–12.7%.

With RO, an increase in flux or conversion rate necessarily increases pressure, as well as TDS concentration. This translates into increased energy consumption. Figure 4.11 shows specific energy consumption as a function of applied flux for MF/UF.

Overall, the energy demand increases with the product of these three parameters. These values are higher than those associated with conventional wastewater treatment, which range from 0.3 to 0.9 $kWh.m^{-3}$, but the water quality obtained is fundamentally different and more appealing.

However, this energy demand is much lower than that of desalination, which totals between 3 and 3.9 $kWh.m^{-3}$, due to the high concentrations of dissolved salts in the feed water (35,000–45,000 $mg.L^{-1}$), in other words, 40–60 times higher than in municipal wastewater. The logic of reusing water using membrane technology as a sustainable alternative to seawater desalination is therefore justifiable.

Depending on the quality of the feed water, a compromise has to be found between conversion rate and clogging. The latter is also a function of permeate flow. Permeability Lp depends on temperature and membrane clogging. The higher Lp is, the greater the membrane's hydraulic performance.

– The flow rate produced (Qp, $m^3.J^{-1}$) determines the number of modules to be used.

– The conversion rate (Y %) is the hydraulic efficiency of the system and represents the ratio between the flow produced (Qp) and the flow at the inlet (Qin):

$$Y\,\% = \frac{Qp}{Qin}$$

It is easy to see that the higher the conversion rate, the lower the concentrate flow rate, and therefore the lower the tangential sweeping effect, and the more likely the membrane is to clog. The conversion rate for a membrane element is 15%. From an economic point of view, this must be as high as possible. From a technical point of view, however, the choice of conversion rate essentially depends on the salinity of the water to be treated.

– The total membrane surface area (*St*) for a train is determined from the water flow rate produced and the flux selected:

$$St\ (m^2) = \frac{Qp}{Jw}$$

– The minimum number of modules per train is obtained by the ratio between the total membrane surface area of a train and the unit surface area of a module (*Sm*):

$$\text{Number of modules/train} = \frac{St}{Sm}$$

– The salt passage rate, which is the ratio between the concentration of the compound in the permeate (*Cp*) and its concentration in the feed water (*Cin*), is as follows:

$$X\ \% = 100\ \frac{Cp}{Cin}$$

– The rejection rate (*R*%), which represents the membrane's ability to reject unwanted salts, is:

$$R\% = 1 - X$$

For certain compounds, such as natural organic substances, pesticides and micropollutants, high rejection is required. For ions, the rejection required depends on the treatment objectives and their initial concentration in the water. For a given membrane, rejection can be influenced by clogging, temperature, hydraulic conditions (flux, conversion rate) and solution composition (coupling effect, etc.).

– The flux of salts (*Js*, $mS.m^{-2}.h^{-1}$), which is a function of the concentration upstream and downstream of the membrane and is independent of the applied pressure, is written as follows:

$$Js = Jw\ (Cin - Cp)$$

4.3.2.3. *Mass flow diagram*

Figure 4.12 shows a mass balance of flows on a membrane module. In terms of flow, the feed flow Qin is the sum of the other two flows (Qp permeate and Qc concentrate):

$$Qin = Qp + Qc$$

In terms of mass flows, we obtain:

$$Qe\ Cin = Qp\ Cp + Qc\ Cc$$

Cin, Cp and Cc are, respectively, the concentrations of salts, ions or organic compounds in the feed water, permeate and concentrate. The conversion rate is given by:

$$Y = Cp/Cin$$

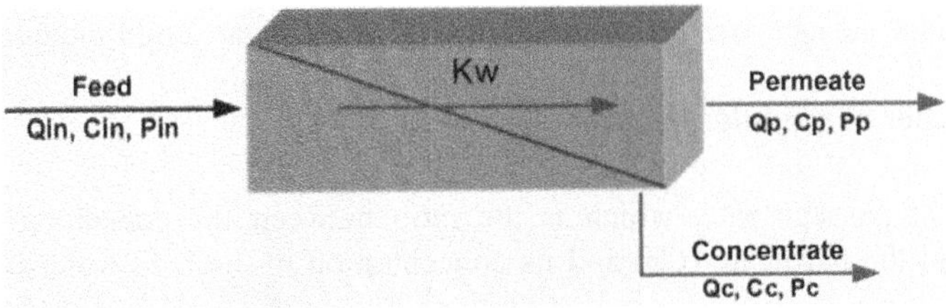

Figure 4.12. *Mass flow diagram*

4.3.2.4. *Concentration factor*

The concentration factor is linked to the conversion rate (Y%) and is an important parameter for system design. The more water recovered as permeate (high conversion rate), the more salts and contaminants are concentrated in the concentrate. This can lead to a saturation potential on the membrane surface when the concentration factor is too high:

$$\text{Concentration factor} = \frac{1}{(1-Y)}$$

For example, with a recovery rate of 75%, the concentration factor is 4. This means that the concentrations of ions and compounds present in the concentrate are four times higher than those in the feed water. Thus, if the feed water consists of 500 ppm TDS, the TDS concentration in the concentrate is 500 × 4 = 2,000 ppm TDS.

4.3.2.5. *Pressure tubes and number of modules per tube*

The number and size of membrane trains must correspond to the requirements of non-binding operation. The choice may be between using several membrane trains with alternating N-1 operation, reduced-capacity membrane trains operating in N, or a limited number of higher capacity trains. The number and size of each train should facilitate start-up and shut-down in order to meet demand. Where appropriate, a train should meet the minimum daily demand. Redundancy can be imposed by the specifications, so as to ensure that the maximum quantity of water produced is met with one train that is out of service and/or one train being cleaned. Often, a larger storage tank is suggested to compensate for lower water production during cleaning or maintenance operations that last less than 8 h.

Determining the number of pressure tubes and the number of elements per tube is generally a process guided by economic objectives, as well as by the limits imposed by inlet and concentrate flow rates.

The number of elements per pressure tube is determined by a manufacturer-specified minimum-maximum concentrate flow rate produced by the last element in the tube. This flow rate depends on the recovery rate and the number of pressure tubes that are in parallel.

An economic assessment of the various options takes into account the capital costs of pressure tubes and associated equipment, as well as the energy costs associated with inlet pressures, which are higher when there is an increased number of elements in a single pressure tube.

Determining the number of pressure tubes involves estimating the number of parallel tubes and the number of stages required to achieve the desired recovery rate.

There are basic rules, but each project must be considered individually:

– For systems operating up to 75% recovery rate, two stages with pressure tubes comprising six elements per tube are common.

– For systems with recovery rates between 76 and 87%, two stages with pressure tubes comprising seven elements per tube are common.

– For systems with recoveries >88%, three pressure tube stages with six elements per tube are common.

When the number of elements per pressure tube is chosen, the minimum number of pressure tubes required can be calculated by:

$$\mathrm{NTP} = \frac{\mathrm{NMT}}{\mathrm{NMt}}$$

where:

– NTP: minimum number of pressure tubes required per train;

– NMT: number of modules per train;

– NMt: number of modules per pressure tube.

4.3.3. *Post-treatment for drinking water supply*

The water's calcium-carbonate balance is profoundly altered by acidification upstream of the membranes, as well as by the membrane's retention of TAC and alkalinity, to a greater or lesser extent. Post-treatment can include many of the single operations that are common to conventional drinking water treatment, such as aeration, degassing, pH adjustment, addition of chemicals to control corrosion, fluoridation and disinfection. When the feed water is acidified, there is a high concentration of CO_2 in the treated water and it requires treatment to restore the calco-carbonic balance.

Treatment generally consists of:

– atmospheric degassing followed by soda injection to give the water a light scaling effect;

– injection of limewater (and CO_2 if necessary) to remineralize and restore the water's calco-carbonate balance;

– an injection of CO_2 or sulfuric acid on the calcium filter.

This practice is not considered when the water produced is then mixed with surface water of sufficient mineral composition.

4.3.4. *Disinfection downstream of membranes*

Lastly, safety disinfection must also be provided. This can be achieved, for example, by UV radiation. In all cases, the residual effect of chlorination must be ensured. In addition, a cleaning station ensures that the membranes are periodically unclogged. This station is then combined with a device for neutralizing the pH of the cleaning solutions. Cleaning will be triggered automatically, either when a head loss threshold is reached, or on a timer, at the operator's discretion. The frequency of cleaning will vary according to the quality of the inlet water.

4.4. Reverse osmosis in the wastewater reuse process

When a municipality wishes to reuse its wastewater, certain uses require more pronounced treatment due to the more stringent quality requirements. In general, wastewater treatment plants include pre-treatment, primary treatment and secondary treatment stages.

By the end of the secondary stage, a large proportion of the solid, organic and inorganic pollutants and metals have been removed or reduced. Secondary treatment can therefore provide an effluent that can be safely discharged, but which would not meet the standards for reuse applications.

A tertiary treatment is then implemented for the following uses:

– agricultural irrigation of edible food crops;

– irrigation of residential and public landscaping;

– structural/non-structural fire-fighting;

– toilet flushing;

– decorative fountains;

– industrial process water that may come into contact with workers;

– industrial or commercial cooling or air-conditioning;

– snowmaking;

– floor and municipal vehicle washing;

– boiler feed water.

Various processes can be used to achieve the quality required for each of these uses. A RO membrane will guarantee very high water quality, justified by its specific applied use (food industry, direct or indirect drinking water, etc.).

Although they are extremely efficient on biologically treated wastewater, RO systems must be combined with an effective pre-treatment system to avoid common problems that can lead to system failure, such as clogging, fouling and scaling.

One of the most effective pre-treatment options is a first membrane barrier. Proven processes include the membrane bioreactor (MBR), or a MF or ultrafiltration (UF) membrane installed downstream of the treatment plant.

The MBR provides high-quality feed water to the RO, minimizes the footprint and cost of civil works and reduces treatment plant downtime, thus cutting operating costs. This process is successfully used for industrial and municipal water reuse. The high level of organic matter that is still present in many biologically treated wastewaters, which is discharged through the RO membrane, concentrated at five to six times its inlet concentration, can clog the membrane, particularly toward the outlet of the RO system. Bacterial growth can also occur, as the organic matter contained in wastewater is an excellent food source for microorganisms, even if the bacterial concentration in the MBR permeate is very low. Calcium phosphate scaling can cause problems with RO systems operating on certain types of wastewater. Scaling can be mitigated by operating at lower water recovery, using an acid or other antiscalant to minimize scaling, or by modifying the operating conditions of the treatment plant to reduce the amount of phosphate in the RO feed.

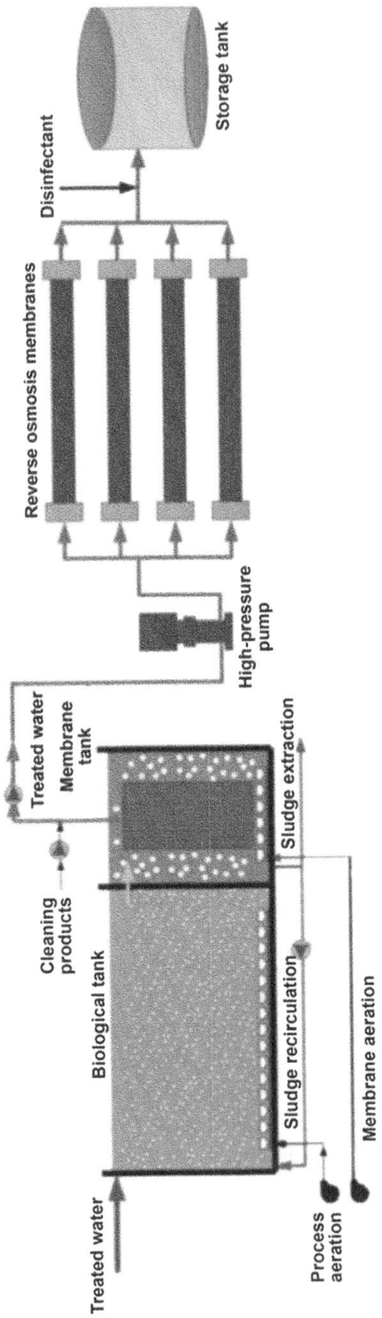

Figure 4.13. *Membrane bioreactor/reverse osmosis system*

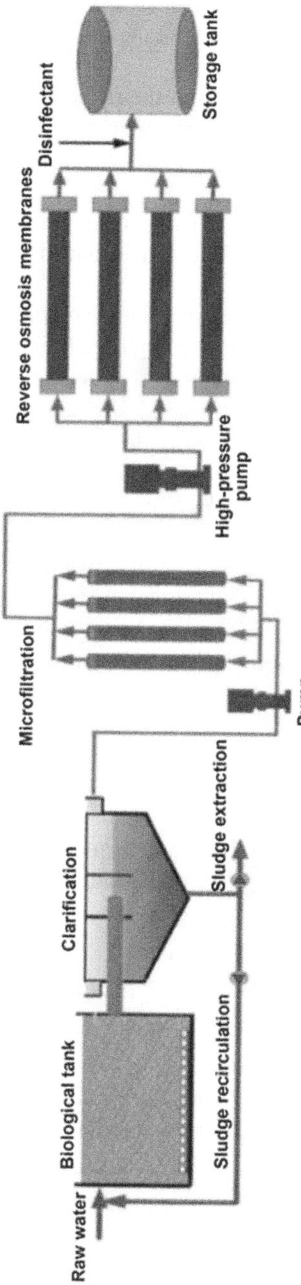

Figure 4.14. *Conventional wastewater treatment plant followed by a double membrane barrier*

These clogging, fouling and scaling problems mean that the RO system has to operate at higher pressures, resulting in increased energy consumption, higher chemical costs for cleaning and shorter membrane life.

These constraints can be minimized, and the overall lifecycle costs of the reuse system can also be reduced. Efficient pre-treatment of RO feed water is the solution, with a first membrane barrier.

When the treatment plant is designed using conventional physical and biological processes, membrane systems can be added in a second stage.

A conventional wastewater treatment plant comprising a biological tank, followed by solids/liquid separation with secondary clarification, may see the outlet effluent undergo further treatment for reuse. The sedimentation process does not remove sufficient bacteria and suspended solids, so sand filtration can be added to improve solids/liquid separation and provide better quality water. However, disruptions in the secondary clarifier can lead to effluents with higher levels of TSS and BOD5, causing mineral clogging and organic fouling of RO membranes.

MF improves the removal of suspended solids as a pre-treatment for RO. Many current water reuse systems use this additional first stage by MF or UF, so as to refine turbidity and greatly reduce microorganism concentrations. These are mostly systems using hollow-fiber membranes, which do an excellent job of providing the RO system with high-quality water.

The addition of an RO system can potentially transform over 80% of wastewater into clean, reusable water.

Not all treatment systems can be implemented in modular units, but RO wastewater treatment systems can be integrated into modular systems, in containers and mobile units. As a result, these simple, easy-to-install units can be added if an increase in flow is required at the treatment plant.

4.5. Performance

RO is recognized as an effective means of providing a safe and reliable water supply for both drinking and non-drinking water. This technology is outstanding in wastewater reuse applications, particularly in drinking water reuse, as it can effectively remove the most harmful organic micropollutants to humans, such as endocrine disruptors, active pharmaceutical compounds and pesticides. It also reduces color, feedwater TOC, salts and metals. Due to the difference in membrane resistance, the pressure driving force and energy costs are lower for LPRO than for

RO applied to desalination. Tightly packed nanofiltration membranes can be used, as they have very little effect on monovalent ions such as chlorides and nitrites.

Table 4.3 shows the physicochemical characteristics of concentrate from a drinking water plant using RO with a recovery rate of 85%.

As a result of their very low cut-off points, LPRO and RO membranes are a well-known barrier for various ions and molecules, with widely proven effectiveness with regard to natural organic substances, organic micropollutants and minerals.

Parameters	Raw water	Permeate	Concentrate
TOC, mg.L^{-1}	5.1	0.23	32
Conductivity at 25 °C µS.cm^{-1}	2,830	230	6,500
Silica, mg.L^{-1}	8.97	1.87	18.80
Calcium, mg.L^{-1}	60	5	363
Magnesium, mg.L^{-1}	20	3	112
Sodium, mg.L^{-1}	15	3	79
Potassium, mg.L^{-1}	4	1	19
Bicarbonates, mg.L^{-1}	84	4	528
Chlorides, mg.L^{-1}	38	2	237
Nitrates, mg.L^{-1}	5.1	2	21
Sulfates, mg.L^{-1}	41	1	264

Table 4.3. *Physicochemical composition of treated wastewater, permeate and concentrate during reverse osmosis treatment*

The information supplied by manufacturers is limited to the membrane's cut-off value (often expressed in daltons), and may take an organic compound, whose molecular weight enables it to be 90% removed, as a reference. Larger molecules are assumed to have a retention of over 90%, while smaller molecules are assumed to have a retention of less than 90%. The separation mechanism and its efficiency depend on the type of membrane (cut-off, morphology, surface electrical charge, structure, etc.), the characteristics of the ions and organic substances (solubility and diffusivity in the membrane), and the temperature. Sorption/diffusion mechanisms contribute to membrane retention and are explained by hydrophobic interactions or hydrogen bonds between organic molecules and the membrane surface.

LPRO is indicated to retain molecules with a molecular weight of 150 Da or more.

Although membrane processes are recognized for producing high-quality permeate for reuse in the drinking water and food industries, there is still some concern about health risks among consumers, as micropollutants are not always completely removed. In this regard, advanced oxidation processes (AOPs) are suggested downstream of the RO membrane train in order to degrade compounds that have escaped exclusion by RO membranes.

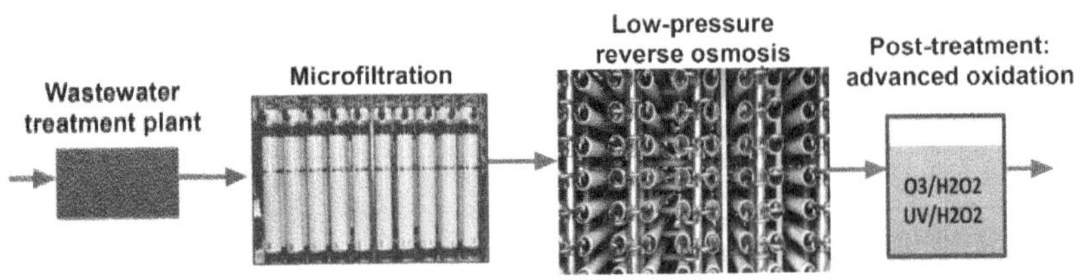

Figure 4.15. *Double-membrane filter with advanced post-treatment oxidation*

AOPs have proven their reliability in wastewater treatment, as they have a high capacity to oxidize almost all organic pollutants, mainly through the generation of hydroxyl radicals (OH·). Their application stands out in the degradation of micropollutants in wastewater, due to their high oxidation potential.

AOP production generally depends on separate or various combination(s) of ozone (O_3), hydrogen peroxide (H_2O_2), catalysts and UV radiation to produce OH· radicals. Ozone-based processes (O_3/H_2O_2, O_3/H_2O_2/UV and UV/H_2O_2) are the most recommended after the RO membrane stage.

Conversely, Fenton-type processes, heterogeneous photocatalysis (TiO_2/UV and TiO_2/H_2O_2/UV systems) and sonochemical and electrochemical AOPs have no applications in municipal environments.

The aim is to fine-tune the quality of the water produced, primarily by targeting residual micropollutants, low molecular weight organic compounds, all pathogens, bacteria and viruses in effluent. The aim is to achieve "zero pathogens".

The disadvantages of AOPs in this configuration include relatively high operating costs and limited large-scale applications.

4.5.1. *Pesticides*

The vast majority of pesticides and their metabolites have molecular weights (MW) generally in excess of 200 Da. As a result, the use of membranes with a cut-off greater than 200 Da is considered sufficient to achieve good removal of these molecules. The efficiency of a LPRO membrane, with regard to pesticides, is based on key parameters such as molecular weight, molecular dimensions (length and width), hydrophobicity/hydrophilicity (log Kow) and log Dn linked to the acid dissociation constant (pKa).

4.5.1.1. *Relationship between pesticide molecular weight and membrane pore diameter*

Generally speaking, pesticide size relative to membrane pore radius is the most important factor determining pesticide release.

The pore size of LPRO membranes is usually smaller than the size of the pesticide (<1 nm). Clearly, the larger the pesticide molecule, the greater the rejection effect of the membrane.

As a result, the molecular weight cut-off (MWCO) of the membranes correlates with the rejection of pesticides of varying molecular weight. MWCO is defined as the molecular weight at which 90% rejection is achieved. Due to the ready availability of molecular weight, it is a parameter that can quickly be used as a performance indicator.

Although molecular weight is not a direct measure of a molecule's dimensions, it still reflects molecular size and is an easily accessible parameter. Figure 4.16 illustrates the correlation between molecular weight and molecule diameter:

$$\text{Molecule diameter (nm)} = 0.065\ (\text{MM})^{0.438}$$

This correlation is valid for the range of MW for which nanofiltration generally intervenes (up to ±600 Da).

The curve (Figure 4.16) can be taken as an initial indicator of the performance of RO membranes with regard to organic micropollutants. However, there may be deviations from this correlation, meaning that other parameters may have an influence. The use of molecular weight does not take into account the spatial geometry of pesticides, which is important when the length and width of pesticides are very different. The rejection of small pesticide molecules by membranes with larger pores can not only be influenced by dimension parameters such as the relative pore size of pesticide and membrane, but also by physicochemical interactions taking place between these organic molecules and the membrane surface.

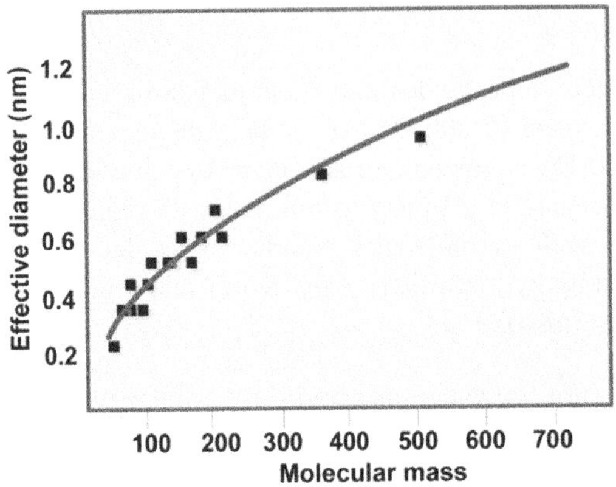

Figure 4.16. *Correlation between molecular weight and molecule diameter*

The porosity of the membrane, the molecular weight of the micropollutants and their dimensions govern the overall retention of micropollutants. This means that a membrane with a sufficiently small pore diameter, such as an RO membrane, can, by steric exclusion, reject all molecules with a molecular diameter greater than that of the membrane. This is the case for most pesticides. This is valid if the molecule is considered spherical.

Membrane	Pore diameter (nm)	Cut-off threshold (dalton)
ESPA2 LD RO (Hydranautics)	0.62	100
XLE (Dow/Filmtec)	0.60	150
BW30 (Dow/Filmtec)	–	100

Table 4.4. *Pore diameters of nanofiltration and low-pressure reverse osmosis membranes*

This may not be sufficient, because although the molecule is not spherical and can penetrate the pores perpendicularly rather than planarly, other physical parameters can predict membrane performance, such as the equivalent width of a molecule, which is defined as the root of the geometric mean between the height of the molecule and its width. By steric exclusion, micropollutant retention increases with equivalent width.

Pesticide	Molecular weight, $g.mol^{-1}$	Effective diameter, nm	Retention % OIBP Toray HMA	Retention % OIBP Dow XLE
AMPA	186.17	0.641	85	85
Atrazine	215.69	0.684	95	95
Desethylatrazine	187.60	0.643	95	95
Deisopropylatrazine	173.60	0.622	95	95
DEDIA	215.68	0.684	94	94
Acetochlor	269.11	0.753	92	92
Acetochlor ESA	315.11	0.807	95	95
Acetochlor OXA	265.13	0.748	95	95
Alachlor	269.17	0.753	95	95
Alachlor ESA	270	0.754	95	95
Alachlor OXA	265	0.748	92	95
Bentazone	240.30	0.717	95	95
Chlortoluron	212.68	0.679	95	95
Diuron	233.1	0.707	95	95
Dimethenamid	275	0.760	95	95
Glyphosate	169.07	0.614	85	85
Heptachlor	373.32	0.869	95	95
Metaldehyde	176	0.625	80	90
Metolachlor	283.60	0.771	95	95
Metolachlor ESA	329.11	0.823	95	95
Metolachlor OXA	279.14	0.765	95	95
Oxadixyl	278.31	0.764	92	92
Simazine	201.66	0.664	95	95
Sulcotrione	328.01	0.822	90	90

Table 4.5. *Retention percentage of Toray HMA and XLE membranes (Dow/Filmtec) as a function of molecular weight and effective diameter of selected pesticides*

Other mechanisms may come into play, such as physicochemical interactions between molecules and the membrane. Ionizable molecules are acids or bases that are characterized by one or more acidity constants (pKa). Depending on the pKa of the pesticide and the pH of the water, pesticides can be neutral, positively charged or negatively charged. The presence of charge-carrying groups in the molecular structure of a pesticide can therefore strongly influence its rejection by the membrane. This explains why the rejection of neutral pesticides is by steric exclusion, while the rejection of positively charged pesticides is by steric exclusion and adsorption to the membrane. Rejection of negatively charged pesticides is by steric exclusion, which overrides the repulsion of negative charges from the membrane.

The presence of natural organic substances (humic substances), as well as various pesticides and other organic micropollutants, can affect the membrane's rejection efficiency. With tight NF membranes (type NF 70 or 90) or LPRO membranes with relatively small pore diameters, rejection may be lower than expected. This is because competitive adsorption between pesticides or competition from humic substances can block the transport of pesticides to the pores.

4.5.2. *Drugs*

Tetracyclines, sulfonamides and hormones have a high adsorption affinity for the RO membrane. Considering the speciation of the molecules, their ionic charge as a function of pH significantly influences their rejection on membranes.

Based on their respective log Kow values, rejection can be estimated. In general, polar compounds are well removed, due to interactions with membrane surfaces.

Size-related exclusion mechanisms result in very high rejections (in other words, over 85%) for carbamazepine, hydrochlorothiazide, propyphenazone and glibenclamide, which have MW above the cut-off threshold of RO membranes. Paracetamol, on the other hand, is only 75% retained, probably due to its small molecular size. However, the concentrations measured in RO water from treated wastewater are so low that these limits are accepted for drinking water.

Removal efficiency considerably decreases for micropollutants with a neutral or positive charge. While RO elimination is close to 100% for naproxen, it is only 20% for acetaminophen (neutral) and 60% for atenolol (positive). The adsorption of ibuprofen by RO membranes is directly linked to the electrostatic repulsion between the pollutant, the membrane and the pH of the solution.

Table 4.6 shows the rejections of HMA and XLE membranes in relation to drugs that are present in water.

In general, the charge effect is greatest when the molecules are much smaller than the pores. Membrane charge can increase or decrease retention, due to electrostatic interactions. When molecules are approximately the same size as pores, charge effects are of little significance, as molecules are mainly retained by a screening effect.

Drug	Molecular mass ($g.mol^{-1}$)	Log Kow	Log D	Molar volume ($cm^3.mol^{-1}$)	Effective diameter (nm)	Equivalent width (nm)	LPRO membrane retention %	LPRO membrane retention %
							Dow XLE	Toray HMA
Acetaminophen (paracetamol)	151	0.46	0.23	120.90	0.704	0.53	80	80
Caffeine	194	–0.07	–0.45	133.30	0.668	0.70	92	92
Carbamazepine	236	2.45	2.58	186.50	0.785	0.73	95	95
17-β-estradiol	272	4.01	3.94	232.60	0.902	0.74	95	95
Bisphenol A	228	3.32	3.86	199.50	0.864	0.79	92	92
17-α-ethynilestradiol	296	3.67	1.48	225.60	0.785	0.85	95	95
Sulfamethoxazole	253	0.89	0.45	173.10	0.850	0.64	95	95
Fenoprofen	242	–0.02	0.37	180.90	0.816	0.83	95	95
Ketoprofen	254	–0.52	0.13	187.90	0.816	0.83	95	95
Naproxen	230	3.18	0.41	192.20	0.802	0.61	95	95
Ibuprofen	206	3.97	0.77	200.30	0.859	0.71	95	95
Gemfibrozil	250	4.77	1.58	221.90	0.997	0.78	95	95
Atenolol	266.3	0.15	2.21	369.8	1.084	0.69	95	95
Bezafribrate	361.8	4.25	1.20	212.8	0.85	–	95	95
Ciproflaxacin	331.3	0.38	0.10	496.7	0.858	–	95	95
Clarithromycin	748	3.16	0.80	–	1.79	–	95	95
Diclofenac	296.1	4.75	0.48	182	0.851	0.900	95	95
DEET	191.2	2.40	1.96	191.5	0.648	0.81	95	95
Erythromycin	733.9	3.06	0.81	611	1.170	0.585	95	95
Estriol	288.4	2.45	2.53	–	0.776	–	95	95
Fluoxetine	309.3	4.05	1.41	412.4	0.801	0.85	95	95
Fenofibrate	360.8	5.20	1.34	644.2	0.857	–	95	95
Fluvoxamine	318.3	2.80	1.75	454.7	0.811	–	96	96
Furosemide	330.7	2.03	–0.82	206.6	0.824	–	92	90
Gabapentin	171.2	–1.10	–1.42	155.1	0.618	–	94	94
Iopamidol	777.1	–2.42	–0.66	–	1.19	–	95	95

Iohexol	821.1	–3.05	2.52	373.2	1.22	–	95	95
Iopromide	791.1	2.49	–0.36	364	1.20	–	95	95
Metformin	129.1	–2.60	–3.24	184.4	0.540	–	85	85
Metoprolol	267.3	1.88	–0.77	477.3	1.127	0.730	95	95
Nonylphenol	220.3	5.76	5.88	231.1	1.083	0.660	95	95
Omeprazole	345.4	2.23	1.80	246.7	0.840	–	98	98
Oxazepam	286.7	2.24	2.80	819.1	0.774	–	95	95
Ofloxacin	361.4	–0.39		240.9	0.857	–	95	95
Progesterone	272.4	4.01	3.83	232.8	0.757	–	95	95
Primidone	218.2	0.91	–0.84	364.2	0.786	0.765	95	95
Propanolol	259.3	3.48	0.21	235.6	0.741	–	95	95
Sotalol	272.3	0.24	–1.70	–	0.936	0.78	95	95
Tramadol	299.8	3.01	1.13	483.5	0.790	–	97	97
Triclocarban	315.6	4.34	5.80	–	0.808	–	95	95
Triclosan	289.5	4.76	5.17	186.7	0.778	0.64	–	–
Trimethoprim	290.3	0.91	0.20	–	0.780	0.88	95	95
Valsartan	495.5	1.49	–0.80	–	0.97	–	95	95
Vanlafaxine	277.4	3.20	0.70	616.4	0.763	–	95	95
Androstenedione	286.4	2.75	2.72	–	0.774	0.74	95	95
Meprobamate	218.2	0.70	0.70	178.8	0.687	0.86	98	98
Testosterone	288.4	3.32	3.00	262.2	0.776	0.63	95	95
Enalapril	376.4	0.07	–1.01	–	0.873	–	98	98
Androsterone	290.4	3.69	3.93	218.3	0.779	0.650	98	98
Salicylic acid	138.1	2.26	–1.20	635.3	0.562	0.260	85	85
Phenacetin	179	1.58	1.68	163	0.809	0.540	75	75
Metronidazole	171	–0.02	–0.27	117.8	0.617	0.660	70	70
Phenazone	188	0.38	0.54	162.7	0.663	0.660	93	93
Estrone	270.4	2.60	3.80	225.3	0.908	0.76	78	78

Table 4.6. *Low-pressure reverse osmosis membrane effectiveness with regard to drugs*

The following equation gives an indication of the membrane's efficiency in rejecting micropollutants such as drugs:

$$R\ (\%) = \frac{100}{1+e(^{-B})}$$

where B is equal to:

$$B = 11.372\ d_{\text{eff}} + 6.795\ l_{\text{eff}} + 107.305\ \text{RS} - 0.399\ \log \text{D} - 116.496$$

d_{eff} is the effective diameter (nm) given by:

$$\text{deff} = \frac{L}{2} + \frac{h}{3.14}$$

where L is the length of the molecule (nm) and h is its height (nm).

l_{eff} is the equivalent width (nm) given by:

$$l\text{eff} = \sqrt{h.l}$$

RS is the salt rejection (%) of the membrane and log D is a measure of log Kow at a given pH for a compound of a certain pKa. This parameter takes the concentration of ionized molecules into account.

4.5.3. *Microorganisms*

RO membranes provide an excellent physical barrier to all microorganisms. They increase the reliability of disinfection for the following reasons:

– The removal achieved on industrial plants remains very good: 6–8 log removal for all germs, particularly for parasites such as *Giardia* and *Cryptosporidium*.

– The water is free of suspended solids and turbidity. Osmosis water has very good UV transmission.

RO has a high removal rate of viruses and bacteria due to its pores being smaller than the size of these microorganisms.

However, this barrier cannot be considered absolute, given the extremely low but not zero risk of water leaks, as well as the limited means of monitoring the plant's integrity.

In the case of spiral-wound modules, leaks can occur at the joints and connectors between the permeate tubes. Leaks may occur during assembly and disassembly, or during hydraulic jolts that cause membranes in the tubes to move.

As module failures no longer guarantee system integrity, post-treatment disinfection can temporarily compensate for the possible presence of microorganisms in the produced water. During repairs, continuous monitoring of the membrane system is essential.

Chlorine can be applied downstream of the membrane, so as to ensure continuous control and a residual effect in the distribution network.

The performance of RO membranes can be summarized as follows:

– significant reduction in pesticides and micropollutants (drug residues, endocrine disruptors);

– reduction in nitrate content by a factor of around 2;

– reduced need for final chlorination compared to a conventional process;

– reduced dissolved organic matter content in water.

4.5.4. *Water losses and energy consumption of membrane processes for water reuse*

Membrane technologies (MF, UF, MBR) offer higher productivity per unit area and limited water losses, with hydraulic efficiencies of over 90%. If we add the hydraulic efficiency of the RO stage, which is around 80–85%, produced water losses for the entire membrane process are between 15 and 20%. Wash water is recycled to the beginning of the treatment plant.

As far as energy consumption is concerned, low-pressure membrane technologies consume more energy than conventional processes. The energy consumption of a MBR ranges from 0.6 to 1.07 $kWh.m^{-3}$.

Hybrid MBR + RO processes have an energy consumption of around 1.9 to 2.8 $kWh.m^{-3}$.

In conclusion, as seen in actual cases of municipal wastewater reuse treatment systems, the wastewater recovery rate and net energy consumption are more likely to be considered the two main indicators of this treatment process.

4.6. Conclusion

The concept and acceptance of wastewater reuse continues to grow, particularly in water-deficient regions and for industrial applications.

For wastewater reuse for drinking water purposes, quaternary treatment technologies (MF, UF, nanofiltration and RO) must be included in the treatment process. Various technologies can be combined, such as membranes, activated carbon and ozonation, in order to enhance the quality of the water produced. Today, membrane technology is one of the most attractive technologies for drinking and non-drinking water reuse applications, with numerous advantages such as efficient pathogen removal and good performance on both micropollutants and mineral matter.

The double membrane process is a high-performance process that ensures high efficiency for the removal of organic and inorganic micropollutants, endocrine disruptors, active pharmaceutical compounds, pesticides, industrial residues and pathogens.

Compared with alternative techniques (activated carbon filtration, O_3/carbon combination, advanced oxidation, etc.), RO membranes offer a better guarantee for drinking water production, as they are effective against a wide range of contaminants.

Membrane filtration has thus become a viable and recognized method of water reuse, producing high-quality water that is safe for human consumption and can be used for a variety of direct and indirect drinking water applications.

Other uses include agriculture, various industries (including food processing), groundwater recharge, salinity intrusion barriers, and commercial and environmental applications.

However, membrane technology still faces limitations in terms of higher investment and operating costs, higher energy consumption and limited membrane service life.

New developments focus on low-energy, low-fouling membranes with high micropollutant retention. The classic TFC membrane has been improved by the development of ultra-thin separation layers. Further developments are underway in the application of nanotechnology to the production of thin-film nanocomposite (TFN) membranes. These new-generation membranes contain aquaporins (AQPs) (protein channels), nanochannels and synthetic carbon nanotubes (CNTs), and graphene oxide (GO).

Ultimately, developments are progressing with a focus on the modification of membranes, using a non-stick surface coating for application in wastewater reuse to limit biological fouling.

4.7. References

Al-Juboori, R.A. and Yusaf, T. (2012). Biofouling in RO system: Mechanisms, monitoring and controlling. *Desalination*, 302, 1–23.

Azais, A., Mendret, J., Gassara, S., Petit, E. (2014). Nanofiltration for wastewater reuse: Counteractive effects of fouling and matrice of the rejection of pharmaceutical active compounds. *Separation and Purification Technology*, 133, 313–327.

Baker, J.S. and Dudley, L.Y. (1998). Biofouling in membrane systems – A review. *Desalination*, 118, 81–89.

Balannec, B., Vourch, M., Rabiller-Baudry, M., Chaufer, B. (2005). Comparative study of different nanofiltration and reverse osmosis membranes for dairy effluent treatment by dead-end filtration. *Separation and Purification Technology*, 42, 195–200.

Bellona, C., Drewes, J.E., Xu, P., Amy, G. (2004). Factors affecting the rejection of organic solutes during NF/RO treatment – A literature review. *Water Research*, 38, 2795–2809.

Ben-David, A., Bason, S., Jopp, J., Oren, Y., Freger, V. (2006). Partitioning of organic solutes between water and polyamide layer of RO and NF membranes: Correlation to rejection. *J. Membr. Sci.*, 281, 480–490.

Blackbeard, J., Lloyd, J., Magyar, M., Mieog, J., Linden, K.G., Lester, Y. (2016). Demonstrating organic contaminant removal in an ozone-based water reuse process at full scale. *Environ. Sci. Water Res. Technol.*, 2, 213–222.

Bouchard, C., Kouadio, P., Ellis, D., Rahni, M., Lebrun, R.E. (2000). Les procédés à membranes et leurs applications en production d'eau potable. *Vecteur environnement*, 3(4), 28–38.

Brandt, D.C., Leitner, G.F., Leitner, W.E. (1993). Reverse osmosis membranes state of the art. In *Reverse Osmosis: Membrane Technology, Water Chemistry, and Industrial Applications*, Amjad, Z. (ed.). Van Nostrand Reinhold, New York.

Buisson, H., Lebeau, T., Lelievre, C., Herremans, L. (1998). Les membranes : point sur les évolutions d'un outil incontournable en production d'eau potable. *L'eau, l'industrie, les nuisances*, 210, 42–47.

Bunani, S., Yörükoglu, E., Yüksel, Ü., Kabay, N., Yüksel, M., Sert, G. (2015). Application of reverse osmosis for reuse of secondary treated urban wastewater in agricultural irrigation. *Desalination*, 364, 68–74.

Chuang, Y.-H., Szczuka, A., Shabani, F., Munoz, J., Aflaki, R., Hammond, S.D., Mitch, W.A. (2019). Pilot-scale comparison of microfiltration/reverse osmosis and ozone/biological activated carbon with UV/hydrogen peroxide or UV/free chlorine AOP treatment for controlling disinfection byproducts during wastewater reuse. *Water Res.*, 152, 215–225.

Dolar, D., Gros, M., Rodriguez-Mozaz, S., Moreno, J., Comas, J., Rodriguez-Roda, I., Barceló, D. (2012). Removal of emerging contaminants from municipal wastewater with an integrated membrane system, MBR–RO. *J. Hazard. Mater.*, 239(240), 64–69.

Dolar, D., Racar, M., Košutic, K. (2019). Municipal wastewater reclamation and water reuse for irrigation by membrane processes. *Chem. Biochem. Eng.*, 3, 417–425.

Dražević, E., Košutić, K., Freger, V. (2014). Permeability and selectivity of reverse osmosis membranes: Correlation to swelling revisited. *Water Res.*, 49, 444–452.

Escobar, I.C., Hong, S., Randall, A.A. (2000). Removal of assimilable organic carbon and biodegradable dissolved organic carbon by reverse osmosis and nanofiltration membranes. *Journal of Membrane Science*, 175, 1–17.

Fan, L., Nguyen, T., Roddick, F.A., Harris, J.L. (2008). Low-pressure membrane filtration of secondary effluent in water reuse: Pre-treatment for fouling reduction. *J. Membr. Sci.*, 320, 135–142.

Greenlee, L.F., Lawler, D.F., Freeman, B.D., Marrot, B., Moulin, P. (2009). Reverse osmosis desalination: Water sources, technology, and today's challenges. *Water Res.*, 43, 2317–2348.

Hofman, J.A.M.H., Noij, T.H.M., Schippers, J.C. (1993). Removal of pesticides and other organic micropollutants with membrane filtration. *Water Supply*, 11, 101–111.

Jacob, M., Li, C., Guigui, C., Cabassud, C., Lavison, G., Moulin, L. (2012). Performance of NF/RO process for indirect potable reuse: Interactions between micropollutants, microorganisms and real MBR permeate. *Desalination*, 46, 75–86.

Kecili, K. (2006). Étude et caractérisation de membranes synthétiques organiques immergées de microfiltration (MF) et d'ultrafiltration (UF) employées pour la production d'eau destinée à la consommation humaine : intensification des opérations de déconditionnement et de nettoyage. PhD Thesis, Université Pierre et Marie Curie, Paris.

Kim, A.S., Chen, H., Yuan, R. (2006). EPS biofouling in membrane filtration: An analytic modeling study. *J. Colloid Interface Sci.*, 303, 243–249.

Lee, K.P., Arnot, T.C., Mattia, D. (2011). A review of reverse osmosis membrane materials for desalination – Development to date and future potential. *J. Membr. Sci.*, 370, 1–22.

Malaeb, L. and Ayoub, G.M. (2011). Reverse osmosis technology for water treatment: State of the art review. *Desalination*, 267, 1–8.

Park, C., Hong, S.-W., Chung, T.H., Choi, Y.-S. (2010). Performance evaluation of pretreatment processes in integrated membrane system for wastewater reuse. *Desalination*, 250, 673–676.

Petersen, R.J. (1993). Composite reverse osmosis and nanofiltration membranes. *J. Membr. Sci.*, 83, 81–150.

Pontié, M., Dach, H., Leparc, J., Hafsi, M., Lhassani, A. (2008). Novel approach combining physico-chemical characterizations and mass transfer modelling of nanofiltration and low pressure reverse osmosis membranes for brackish water desalination intensification. *Desalination*, 221, 174–191.

Potts, D.E., Ahlert, R.C., Wang, S.S. (1981). A critical review of fouling of reverse osmosis membranes. *Desalination*, 36, 235–264.

Racar, M., Dolar, D., Špehar, A., Košutic, K. (2017). Application of UF/NF/RO membranes for treatment and reuse of rendering plant wastewater. *Process Saf. Environ. Prot.*, 105, 386–392.

Radjenovic, J., Petrovic, M., Ventura, F., Barceló, D. (2008). Rejection of pharmaceuticals in nanofiltration and reverse osmosis membrane drinking water treatment. *Water Res.*, 42, 3601–3610.

Shanmuganathan, S., Vigneswaran, S., Nguyen, T.V., Loganathan, P., Kandasamy, J. (2015). Use of nanofiltration and reverse osmosis in reclaiming micro-filtered biologically treated sewage effluent for irrigation. *Desalination*, 364, 119–125.

Sharif, A.O, Merdaw, A.A., Al-Bahadili, H., Al-Taee, A., Al-Aibi, S., Rahal, Z., Derwish, G.A.W. (2009). A new theoretical approach to estimate the specific energy consumption of reverse osmosis and other pressure-driven liquid-phase membrane processes. *Desalination and Water Treatment*, 3(1–3), 111–119.

Shipers, J.C. and Verdouw, J. (1980). The modified fouling index, a method for determining the fouling characteristics of water. *Desalination*, 32, 137–148.

Shon, H.K., Vigneswaran, S., Kim, I.S., Cho, J., Ngo, H.H. (2006). Fouling of ultrafiltration membrane by effluent organic matter: A detailed characterization using different organic fractions in wastewater. *J. Membr. Sci.*, 278, 232–238.

Silva, L.L.S., Moreira, C.G., Curzio, B.A., da Fonseca, F.V. (2017). Micropollutant removal from water by membrane and advanced oxidation processes – A review. *J. Water Resour. Prot.*, 9, 411–431.

Snyder, S.A., Adham, S., Redding, A.M., Cannon, F.S., De Carolis, J., Oppenheimer, J., Wert, E.C., Yoon, Y. (2007). Role of membranes and activated carbon in the removal of endocrine disruptors and pharmaceuticals. *Desalination*, 202, 156–181.

Tang, C.Y., Yang, Z., Guo, H., Wen, J.J., Nghiem, L.D., Cornelissen, E. (2018). Potable water reuse through advanced membrane technology. *Environmental Science and Technology*, 52(18), 10215–10223.

Taylor, J.S. and Jacobs, E.P. (1996). Reverse osmosis and nanofiltration. In *Water Treatment Membrane Processes,* Malleviale, J., Odendaal, P.E., Wiesner, M.R. (eds). McGraw-Hill, New York.

Van der Bruggen, B., Schaep, J., Wilms, D., Vandecasteele, C. (1999). Influence of molecular size, polarity and charge on the retention of organic molecules by nanofiltration. *Journal of Membrane Science*, 156, 29–41.

Warsinger, D.M., Chakraborty, S., Tow, E.W., Plumlee, M.H., Bellona, C., Loutatidou, S., Karimi, L., Mikelonis, A.M., Achilli, A., Ghassemi, A. (2018). A review of polymeric membranes and processes for potable water reuse. *Prog. Polym. Sci.*, 81, 209–237.

Wijmans, J.G. and Baker, R.W. (1995). The solution-diffusion model: A review. *J. Membr. Sci.*, 107, 1–21.

Wintgens, T., Melin, T., Schäfer, A., Khan, S., Muston, M., Bixio, D., Thoeye, C. (2005). The role of membrane processes in municipal wastewater reclamation and reuse. *Desalination*, 178, 1–11.

Yang, Y., Pignatello, J.J., Ma, J., Mitch, W.A. (2016). Effect of matrix components on UV/H_2O_2 and UV/$S_2O_8^{2-}$, advanced oxidation processes for trace organic degradation in reverse osmosis brines from municipal wastewater reuse facilities. *Water Res.*, 89, 192–200.

Wastewater reuse examples

Apostolidis, N., Hertle, C., Young, R. (2011). Wastewater reuse in Australia. *Water*, *3*, 869–881.

Bartels, C.R. (2003). Reverse osmosis membranes for wastewater reclamation. In *IDA World Congress on Desalination and Water Reuse*, Bahamas.

Boake, M.J. (2006). Wastewater reuse: Cases of Gerringong & Gerroa. *Desalination*, 188(1/5), 14–17.

Boake, M.J. and Ovens, G. (2003). Gerringong Gerroa sewage scheme. *Water*, 30(7), 58–61.

Byrne, E., Fitzgibbon, K., Minescu, A., Blanke, J. (2016). Water reuse in the context of the circular economy. Report No. 293, Government of Ireland, Wexford.

Council for Environmental Protection and Heritage (2009). National Council for Health and Medical Research and Ministerial Council for Natural Resource Management. Report, Australian Guidelines for Water Recycling, Canberra.

Department of Sustainability, Environment, Water, Population and Communities (2004). Intergovernmental agreement on the national water initiative. Report, Canberra.

Du Pisani, P.L. (2005). Surviving in an arid land: Direct reclamation of potable water at Windhoek's Goreangab Reclamation Plant. *AridLands*, 56, 62–73.

Foley, J., Batstone, D., Keller, J. (2006). Water recycling in Australia – Position paper. Report, Advanced Water Management Center, Brisbane.

Gaid, K., Bonne, P., Gisclon, A. (2002a). Wastewater reuse: The case of Durban. In *Paper Presented at the Biennial Conference of the Water Institute of Southern Africa (WISA)*, WISA, Durban.

Gaid, K., Bonne, P., Gisclon, A. (2002b). The Durban water recycling project – The vision becomes reality. In *Paper Presented at the Biennial Conference of the Water Institute of Southern Africa (WISA)*. WISA, Durban.

Gisclon, A., McCarley, S., McNally, K. (2002). The Durban water recycling project – The vision becomes reality. In *Paper Presented at the Biennial Conference of the Water Institute of Southern Africa (WISA)*. WISA, Durban.

Haarhoff, J., Van der Merwe, B. (1996). 25 years of wastewater reuse in Windhoek (Namibia). *Water, Science and Technol.*, 33(10/11), 25–35.

Hopkins, L. and Barr, K. (2002). Operation of a water reclamation plant to convert wastewater effluent into high quality water for industrial reuse. In *Proceedings of the 3rd IWA World Water Congress*, Melbourne.

Jacobsen, M., Webster, M., Vairavamoorthy, K. (2012). *The Future of Water in African Cities: Why Waste Water?* World Bank, Washington, D.C.

Lee, H. and Tan, T.P. (2016). Singapore's experience with reclaimed water: NEWater. *International Journal of Water Resources Development*, 32(4), 611–621.

Maquet, C. (2020). Wastewater reuse: A solution for the future, field actions science reports. *Journal of Field Actions*, 22, 2020–2028.

Menge, J. (2006). Treatment of wastewater -reuse in drinking water system of Windhoek. In *Water Institute of Southern Africa Conference*. WISA, Namibia.

Ng, P.J.H. (2018). Singapore: Transforming water scarcity into a virtue. In *Assessing Global Water Megatrends*, Biswas, A.K., Tortajada, C., Rohner, P. (eds). Springer, Singapore.

Queensland Government, Environmental Protection Agency (2005). Queensland water recycling guidelines. Report, Brisbane, 84.

Queensland Water Commission (2008). Water for today, water for tomorrow–South East Queensland Regional Demand Management Program 2007–2009. Report, Brisbane.

Solley, D., Boerlage, S., Traves, W. (2010). Climate resilient water supply in southeast. *Water*, 5, 251–263.

Water Utilities Association of Australia (2010). WSAA review 2009–2010. Urban Water Industry Performance and Future Projections. Report, WUAA, Melbourne.

Western Water Canada (2022). *Proceedings of the Western Canada Water Conference*. Western Water Canada, Calgary.

Zhang, X., Cahill, D.G., Coronell, O., Mariñas, B.J. (2009). Absorption of water in the active layer of reverse osmosis membranes. *J. Membr. Sci.*, 331, 143–151.

5

Applications for Drinking Water, Specific Industrial Water and Groundwater Recharge

5.1. Windhoek: drinking water, specific industries and groundwater recharge

The southern African country of Namibia is famous for its vast desert landscapes. It is one of the driest countries in the world. Crossed by the Kalahari and Namib deserts, with the Atlantic coast at its western border, there is virtually no rainfall. The capital, Windhoek, is under constant water stress, with an average annual rainfall of barely 300 mm, and aquifers that are prevented from recharging as a result of high evaporation rates.

Windhoek mainly relies on water supply from boreholes and three surface dams in intermittent rivers that are 60–200 km apart. Strong population growth and industrial development have led to a significant demand for water. Existing resources are very limited and groundwater is difficult to replenish due to irregular rainfall, leading to a critical situation. As a result, water scarcity is a harsh reality for many Namibians, including the 400,000 inhabitants of the capital, which is the hub of the country's industrial sector.

Namibia has also been a pioneer in wastewater recycling since the 1960s. In 1968, a system for the direct recovery of water from domestic wastewater was launched in Windhoek in order to supplement the city's drinking water supply. This system has undergone a succession of modifications and improvements over the years.

For a color version of all figures in this chapter, see www.iste.co.uk/gaid/rcusc2.zip.

In 1995, the dams that were supplying Windhoek were almost empty and the city faced an extreme crisis. It was even announced that the city's taps would run dry over the next six months.

The authorities turned their attention to finding new, non-conventional resources. While the notion of recycling wastewater into drinking water was not well accepted in many countries, Namibia had no choice. In the absence of other available options, the inhabitants accepted anything that could reduce the water shortage.

Recognizing that wastewater is a potential resource, the authorities decided to use it as an alternative to offset their deficit. As early as 1998, the construction of a new plant to reuse wastewater as drinking water was planned.

Figure 5.1. *Windhoek wastewater reuse plant (Namibia)*

The Gammams wastewater treatment plant (GWTP) receives most of the domestic effluent, and includes biological treatment with nitrification and chemical dephosphatation. It supplies the new Goreangab wastewater reuse plant, with an output of 21,000 $m^3.day^{-1}$. Another source from the Goreangab dam can be mixed in whole, or in part, with the effluent from the treatment plant.

The various stages involved in the reuse plant include:

– raw water pre-ozonation. In the event of temporary ozonation failure, powdered activated carbon replaces ozone;

– coagulation with $FeCl_3$, flocculation and flotation with dissolved air. Hydrochloric acid can be added if required to control pH;

– dissolved air flotation (DAF);

– dual-media filtration: anthracite-sand with soda and $KMnO_4$ dosing to treat iron and manganese residues;

– intermediate ozonation: ozone is dosed at three dosing points, with ratios of 1–1.5 mg O_3.mg^{-1} DOC (dissolved organic carbon) and a contact time of 15–20 min;

– at the reactor outlet, H_2O_2 is added to remove all residual ozone;

– biological activated carbon (BAC) filters;

– granular activated carbon (GAC) filters: the overall contact time of the two GAC filters is 30 min;

– ultrafiltration (UF);

– disinfection by chlorination: break-point chlorination to ensure a free chlorine residual of 1 mg.L^{-1}, subject to a contact time of 60 min;

– stabilization with the addition of caustic soda to regulate pH and ensure a calcium carbonate precipitation potential (CCPP) of 4 mg.L^{-1}.

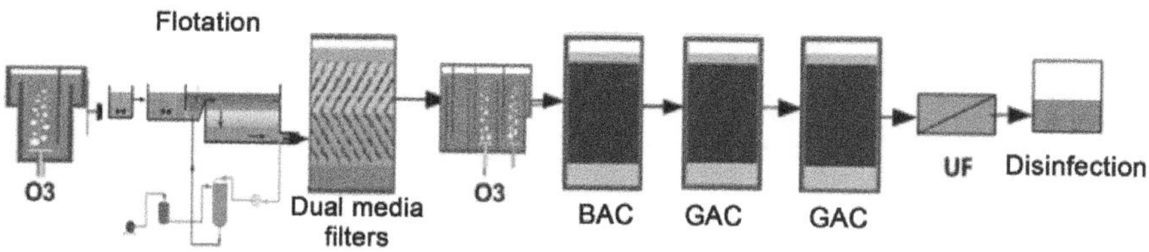

Figure 5.2. *Windhoek wastewater reuse plant process*

The plant is fully automated, with in-line instrumentation in all the main treatment units for continuous monitoring. Produced water is pumped to the New Western plant, where it is blended with another source of drinking water in a 1/3 ratio. Water quality monitoring at the Goreangab plant is very stringent, so as to ensure the quality of the water delivered at all times, as well as to uphold consumer trust.

Analyses include detection of bacteria, viruses, protozoa, helminth eggs, virology, organic carbon, pharmaceutical substances, endocrine disruptors, disinfection by-products, pesticides, algal toxins, taste and odor compounds.

Figure 5.4 shows the plant's performance with regard to residual organic carbon, COD and UV-254 along the treatment process. Residual organic concentrations are as follows: DOC: 2.82 mg.L^{-1}; COD: 11 mg.L^{-1}; UV-254: 0.027 abs.cm^{-1}.

a) b)

Figure 5.3. *Flotation unit a) and ultrafiltration unit b)*

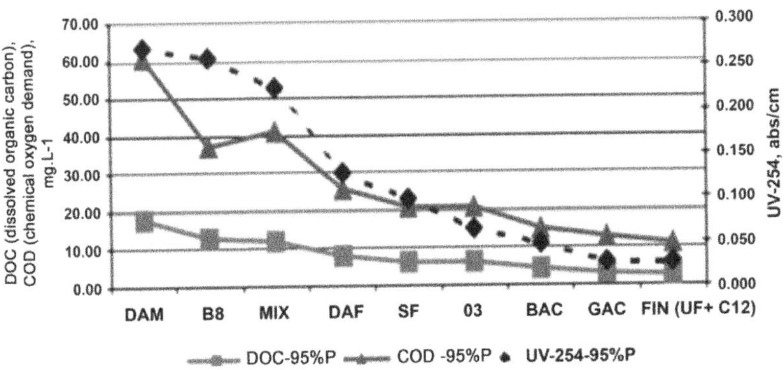

Figure 5.4. *Plant performance on organic substances expressed in DOC, COD and UV-254*

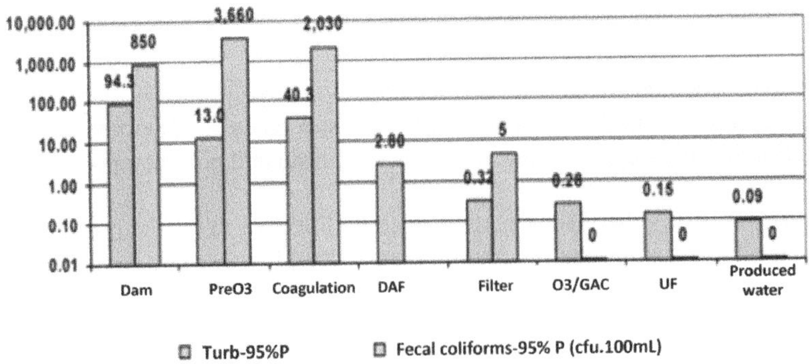

Figure 5.5. *Changes in turbidity and fecal coliforms throughout the process*

Figure 5.5 shows the results of particle (turbidity) and bacterial (fecal coliform) analyses.

Turbidity in the final water is 0.09 NTU. Bacteria are well removed during the interozonation stage.

Parameters	Plant inlet	Water produced	Contractual guarantee
Turbidity NTU	15–52	<0.10	0.10
pH	7.5–8.5	7.5–8.0	–
Color, mg.L^{-1} Pt/co	30–50	<5	–
Conductivity, µS.cm^{-1}	64–68	75–85	150
COD, mg.L^{-1}	50–80	8–11	–
TOC, mg.L^{-1}	15–20	<2	5
Fe	1–2.4	0.01–0.03	0.10
Mn	0.04–0.08	0.005	0.05
Al	–	0.10	
N-NH_3	1–3	0.05	0.10
THM	–	<20	–
Alkalinity, mg $CaCO_3$.L^{-1}	112–125	80–95	–
Hardness, mg $CaCO_3$.L^{-1}	120–150	120–135	–
Chlorides, mg.L^{-1}	80–98	130–150	250
Sulfates, mg.L^{-1}	42–50	42–50	200
Phenols, mg.L^{-1}	0.4–0.5	0–0.02	0.1
Formaldehydes, mg.L^{-1}	0.1–0.3	0	0.10
Chlorophyll, µg.L^{-1}	–	<1	1
UV 254 nm	0.32–0.60	0.001–0.008	0.006

Table 5.1. *Physicochemical quality of plant inlet water and produced water*

The average quality of input and output water is summarized in Table 5.1, which shows that the required guarantees are met on all parameters.

Parameters	Treatment plant outlet	Treatment plant output	Analyzed
Aerobic germ count at 22°C, u/mL	6.5×10^3–5.7×10^5	80	0–1
Total coliforms, CFU/100 mL	2.45×10^5	0	0
Fecal coliforms, CFU/100 mL	2.2×10^4	0	0
***Escherichia coli*, CFU/100 mL**	10^4	0	0
Coliphages, u/100 mL	–	0	0
Enteroviruses, u.10 L^{-1}	–	0 or 4 log	0
***Pseudomonas aeruginosa*, u/250 mL**	2—200	0	0
Fecal streptococci, CFU/100 mL	270	0	0
***Clostridium* sp., CFU/100 mL**	1.1×10^4	0 or 6 log	0
***Cryptosporidium*, u.100 L^{-1}**	–	0 or 6 log	0
***Giardia lamblia*, u.100 L^{-1}**	–	0 or 6 log	0

Table 5.2. *Bacteriological analysis results compared with required guarantees*

The drinking water produced by the Goreangab facility is constantly subjected to quality controls, not just to ensure the safety of the sources but also to earn the public's trust. In addition, there is a health risk management program linked to research projects. It includes advanced tests in terms of virology, parasitology, toxicity, pesticides, algal toxins, etc., carried out by external laboratories. All these efforts have paid off: the capital has been able to significantly increase its water resources. The Namibian authorities have a motto: "In Windhoek, every drop of water counts".

Practical experience in Windhoek demonstrates that a direct wastewater reuse system is a practical and responsible way of increasing the supply of drinking water in arid regions, but that it requires comprehensive planning, ongoing training and commitment for its continued success. The water produced has led to the installation of new water distribution points in the towns, helping to improve the health and safety of local residents.

The Windhoek wastewater treatment plant has become a global reference and a model for innovative and sustainable water management. It is also a successful example of public/private partnership (PPP), increasingly visited by officials from all over Africa, as well as numerous experts from Australia, Singapore and the United States.

5.2. The Durban plant (South Africa)

5.2.1. *For industrial use requiring rigorous quality*

South Africa is a water-poor country. Durban, the country's second largest city and one of its main commercial centers, depends on the availability of water resources for its sustainable economic and social development. In the 1990s, Durban was faced with a constraint on the sewage treatment plant's capacity. The existing infrastructure could not cope with the city's population growth and economic development. The municipality had to invest in new infrastructure and increase wastewater collection to avoid negative impacts on its citizens and the environment. Durban's first option was to invest in a new marine outfall pipeline. However, infrastructure costs were very high and the city considered alternative solutions. Through a PPP, the municipality successfully implemented a project to recycle wastewater for industrial use.

The concept of tapping the potential of municipal and industrial wastewater led to the emergence of wastewater reclamation, recycling and reuse as integral components of city water resource management. The inherent benefits associated with reclaiming treated wastewater included the preservation of water resources, environmental protection and economic benefits. In Durban, one of the catalysts for this strategy was the provision of alternative solutions to meet industrial needs. Water shortages, particularly during periods of drought, had caused strict controls on water consumption rates.

The existing treatment plant is an activated sludge plant, and bidders were asked to investigate the possibility of installing tertiary treatment and selling the water produced to nearby industries, which included a paper mill, a refinery and a chemical complex. Thus, instead of increasing the capacity of the existing marine outfall to discharge primary wastewater into the ocean, Durban explored the possibility of treating it further and reusing it for industrial purposes. Mondi, a paper industry, and Sapref, an oil refinery, expressed interest in receiving the treated wastewater. The aim of the project was to treat approximately 47,500 $m^3.day^{-1}$ (around 10% of the city's wastewater) to a quality that was acceptable for industrial reuse: 85% of the treated water would go to Mondi and the remainder to Sapref.

In order to supply recycled water to the two industrial users, Ethekwini Water Services needed to upgrade the existing activated sludge process, build a tertiary treatment system, increase tank storage and install a reticulation system for the reclaimed water. One of the complexities of the project was that Mondi needed high-quality water to produce fine paper.

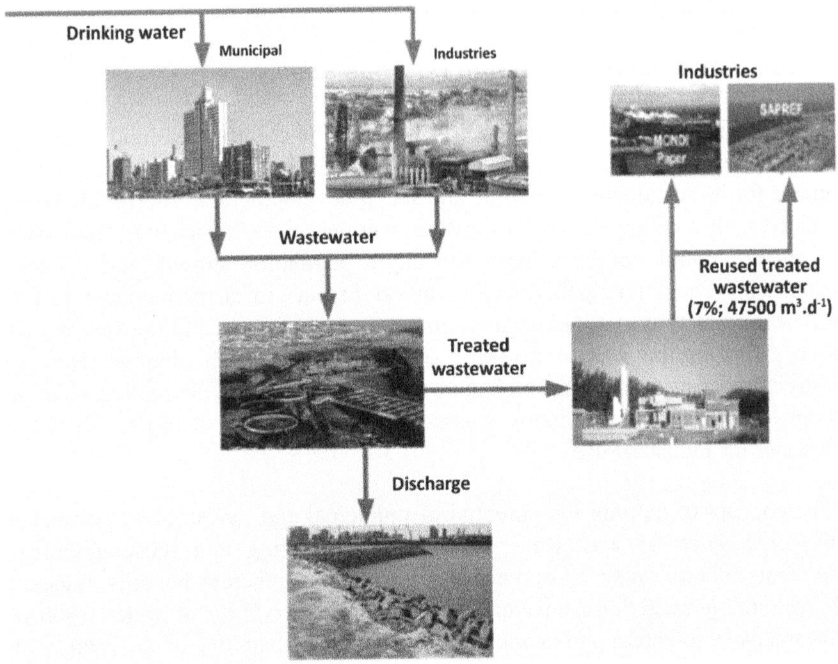

Figure 5.6. *Diagram of wastewater reuse in Durban (South Africa)*

The Durban project was a BOOT project (build, own, operate, transfer, 20 years of operations) consisting of:

– improved management of the existing Umlaas urban wastewater treatment plant;

– reuse of treated wastewater for industrial applications.

The existing activated sludge plant treats an average of 43,000–48,000 m^3/day of mixed urban and industrial wastewater. Due to the daily profile and low periodic inflow capacity, it has been necessary to use industrial wastewater from the Jacobs site (a group of industries located nearby) as a back-up, so as to ensure that the treatment process is able to produce the required quality and flow rate.

The plant's overall treatment includes biological treatment, followed by a physicochemical treatment used as tertiary treatment, which constitutes the wastewater reuse stage.

Figure 5.7. *Durban wastewater treatment plant*

Figure 5.8. *Wastewater treatment plant secondary clarifier*

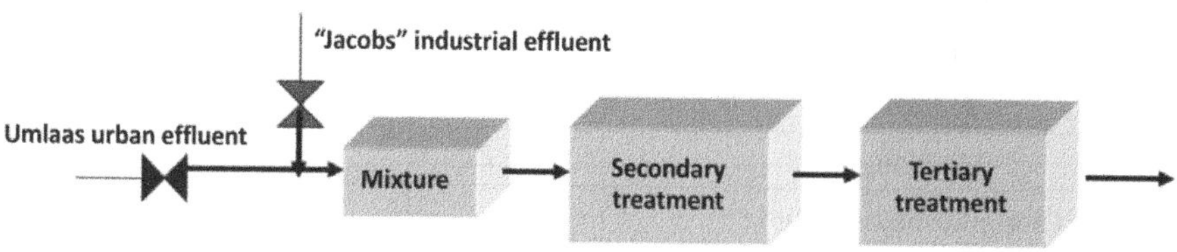

Figure 5.9. *Use of Jacobs industrial sewer to compensate for Umlaas flow deficit*

Parameters	Average	95% P	Base values
Temperature, °C	21.6	19.3	18
pH	7.1	7.4	6.8–8
Alkalinity, mg.L^{-1} $CaCO_3$	209.4	229	250
Total hardness, mg.L^{-1} $CaCO_3$	66	78.1	130
Calcium, mg.L^{-1}	16.3	19.3	–
Magnesium, mg.L^{-1}	6.0	7.1	–
Sodium, mg.L^{-1}	73.2	82.9	90
Potassium, mg.L^{-1}	11.7	14.5	15
Iron, mg.L^{-1}	1.2	1.4	1.5
Manganese, mg.L^{-1}	0.10	0.12	0.15
Silica, mg.L^{-1}	4.3	4.8	5
Chlorides, mg.L^{-1}	75.1	81.5	85
Sulfates, mg.L^{-1}	22.0	29.0	80
Aluminum, µg.L^{-1}	230.9	298.5	1,000
Zinc, mg.L^{-1}	0.2	0.3	0.6
Lead, µg.L^{-1}	19.6	31.8	100
Cadmium, µg.L^{-1}	1.7	2.4	50
Chromium, µg.L^{-1}	21.6	31.9	100
COD (filtered), mg O_2.L^{-1}	599.1	687.5	320
BOD5, mg.L^{-1}	298.5	386.9	–
Color, °Hazen	56.8	72	80
Conductivity, mS.m^{-1}	71.5	77	–
TDS, mg.L^{-1}	425.8	455.8	480
Nitrates, mg.L^{-1}	0.3	0.4	0.5
Nitrite, mg.L^{-1}	0.3	0.3	0.5
Ammonia, mg.L^{-1}	23.1	27.5	32
NTK, mg.L^{-1}	41.7	50.5	55
TSS, mg.L^{-1}	239.3	307.2	350
MVS, mg.L^{-1}	75.2	77.6	–
Total phosphates, mg.L^{-1}	7.6	8.4	10
Oils and fats, mg.L^{-1}	30.6	51.1	–

Table 5.3. *Composition of raw wastewater from Umlaas treatment plant*

This solution required the following implementations:

– use of maximum aerator capacity (to treat flows with high BOD5 and N-NH_3 loads);

– improved sedimentation conditions with the installation of a fourth clarifier;

– addition of ferric chloride to the activated sludge tank to remove excess phosphates;

– provide additional oxygen in biological tanks to improve removal of pollutants;

– implement special automated management procedures for industrial needs.

5.2.2. *The wastewater reuse plant*

The treatment technology was conceived as a combination of fairly conventional processes and an innovative combination of different stages. The treatment process provided a unique project and guaranteed that the recycled water would meet the rigorous water quality requirements of the industrialists (Mondi and Sapref). Indeed, the required characteristics of reused water are very close to those of drinking water, or even better, for certain parameters.

5.2.3. *Tertiary treatment sizing*

The tertiary treatment process is shown in Figure 5.10.

5.2.3.1. *Coagulation-flocculation-settling (Multiflo, Veolia)*

The aim of this first processing stage is to:

– precipitate phosphates;

– reduce overall water conductivity by precipitating carbonates;

– remove suspended solids from clarifiers;

– precipitate any metals present.

The Multiflo includes a coagulation zone, a flocculation zone and a settling zone. The settling zone is equipped with inclined lamellae (60°) made of food-grade polyvinyl chloride with a prismatic profile. Two lamellae, back to back, thus form a module made up of hexagonal-section profiles. The flocculated water is then distributed beneath the modules through an orifice that extends across the entire width of the unit. The water circulates up and down between the lamellae in the opposite direction (countercurrent) to the flocs, which settle on the plates and slide downwards by gravity.

Two compromises are sought in the lamellar zone:

– optimized lamella height;

– evacuation of deposited floc by gravity flow and use of a sufficiently low angle of inclination.

To prevent flocs from shearing off when water currents interact with flocs, the lamellae are placed sufficiently close together (45 mm), so as to encourage flocs to slide toward the clarifier.

The effective settling area of a lamellar module is equal to its horizontal projected area, which is defined by:

$$Sd = N \times L \times l \times \cos(\alpha)$$

where:

– N is the number of lamellae;

– L is the length of the lamellae;

– l is the width of the lamellae;

– α is the angle of inclination of the lamellae.

Partial decarbonation can be achieved by injecting lime into the Multiflo according to the following chemical reactions:

$$Ca(OH)_2 + Ca^{2+} + 2HCO_3^- \rightarrow 2CaCO_3 + 2H_2O$$

$$Ca(OH)_2 + Mg^{2+} + 2HCO_3^- \rightarrow CaCO_3 + MgCO_3 + 2H_2O$$

Sizing is based on a Hazen velocity of 1.5 $m.h^{-1}$ for 44,000 $m^3.day^{-1}$.

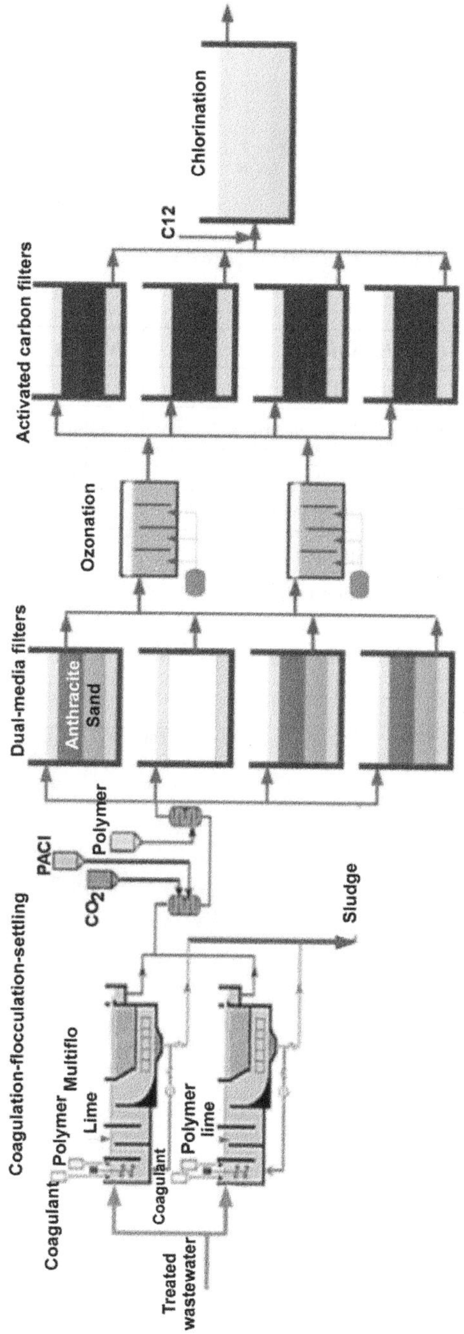

Figure 5.10. *Tertiary treatment process*

Figure 5.11. *View of the wastewater reuse plant*

Parameters	Values
Number of Multiflo	2
Maximum flow rate, $m^3.h^{-1}$	2,048
Unit flow rate, $m^3.h^{-1}$	1,024
Mirror surface area (unit), m^2	60
Hazen speed, $m.h^{-1}$	1.5
Mirror speed, $m.h^{-1}$	17

Table 5.4. *Clarifier operating characteristics*

Figure 5.12. *Multiflo clarifiers*

The settled water flows to the dual-media filters. A static mixer installed upstream of the filters vigorously mixes the coagulant (Al polychloride) injected at the outlet of the lamellar clarifier.

To regulate pH between 6.5 and 7, CO_2 is injected through porous tubes within the same static mixer, promoting coagulation of residual colloids. The main advantage of carbonic acid over hydrochloric or sulfuric acid is that no ions are added to the water. As a result, conductivity does not increase.

The flocs generated by the coagulant trap residual turbidity within the filters and enable residual phosphates to be precipitated.

Number of filters	U 4
Total filtration area (m²)	200
Filtration rate (m/h)	9.16
Unit surface area (m²)	50
Layer heights (m): **– Anthracite: TE: 1.4 mm** **– Sand: TE: 0.8 mm**	 1.00 0.50

Table 5.5. *Sizing dual-media filters*

Four dual-media filters (anthracite/sand) carry out the first filtration stage.

The presence of anthracite in the upper part of the filter media helps retain a greater volume of flocs, while ensuring filtration cycles of over 24 h. The filter floor is specially designed to meet the essential role it must play in filter operation and backwash efficiency.

5.2.3.2. *Interozonation*

The filtered water, freed of turbidity, then moves on to the additional treatment stage, which combines ozone and activated carbon. Ozone is the most powerful oxidant available on the market. It offers the best available technology for oxidizing organic matter and eliminating color.

Powerful ozone generators (Siemens supplier) are installed to oxidize refractory organic carbon concentrations. The ozonation system is capable of producing 12 kg O_3 per hour, at a maximum dosage of 6 g.m^{-3}, with a performance transfer of 85%. Contact time in each oxidation chamber is 8 min.

Figure 5.13. *Ozone generator (Siemens (ex-Trailigaz))*

Figure 5.14. *Granular activated carbon filters*

5.2.3.3. *Granular activated carbon filters*

GAC filtration (Jacobi, ex-Pica supplier) is a biological filtration process for the presence of biodegradable organic carbon from the previous interozonation process. Biological activation ensures activated carbon longevity, estimated at around 5 years.

Number of filters (u)	4
Unit filtration area (m²)	40
Filtration rate on four filters (m.h^{-1})	11.4
Filtration rate on three filters (with one filter being washed) (m.h^{-1})	15.2
Material height (m)	2.10
EBCT (min)	11.0

Table 5.6. *Sizing granular activated carbon filters*

Final disinfection is the last stage in the treatment process. It is carried out using sodium hypochlorite, in a reaction chamber with a contact time of 30 min.

5.2.3.4. *Sludge disposal*

Sludge extracted from clarifiers, dual-media filter backwashes and GAC contactors is conveyed to a 500 m^3 sludge tank. It is recirculated to the inlet of the treatment plant.

5.2.3.5. *Results*

Reused water qualities are summarized in Table 5.7. They comply with the guarantees required by industrial clients.

5.2.3.5.1. Turbidity

Turbidity is one of the main parameters required by industry. Produced water turbidity ranges from 0.2 to 0.5 NTU, with a median value of around 0.4 NTU and a guaranteed value of 1 NTU. The triple barrier (clarifier, dual-media filter, GAC) justifies the low turbidity obtained and the 100% compliance for this parameter.

Parameters	Unit	Guaranteed value	Compliance (%)
pH	–	6.5–7.5	100
Conductivity	µS cm^{-1}	<600	95
Color	Hazen	<5	100
Turbidity	NTU	<1	100
TSS	mg.L^{-1}	<5	100
Chlorine residual	mg.L^{-1}	0.2–0.6	100
Ammonia as N	mg.L^{-1}	<1	100
Total hardness	mg.L^{-1}	<130	94
Alkalinity	mg.L^{-1}	<100	94
TDS	mg.L^{-1}	<500	–
Nitrate	mg.L^{-1}	<10	100
Nitrite	mg.L^{-1}	<1	100
Chlorides	mg.L^{-1}	<90	90
Iron	mg.L^{-1}	<0.04	98
Sulfate	mg.L^{-1}	<90	80
Sodium	mg.L^{-1}	<90	56
Potassium	mg.L^{-1}	<15	100
Silica	mg.L^{-1}	<12	100
Aluminum	mg.L^{-1}	<0.4	100
Manganese	mg.L^{-1}	<0.05	100
Phosphate	mg.L^{-1}	<0.4	98
Oils and fats	mg.L^{-1}	<5	100
DOC	mg.L^{-1}	<3	100
Copper	mg.L^{-1}	<0.01	100

Table 5.7. *Performance and compliance of different parameters of produced water*

5.2.3.5.2. Color

The final color obtained ranges from 2 to 4 °H, with a median value of around 3 °H. This result is due to the coagulation-flocculation stage and the ozone/GAC pair. Compliance is 100% for this parameter.

5.2.3.5.3. Alkalinity

Alkalinity measured in mg of $CaCO_3$ per liter has an average value of 90, with a maximum value of 120. The variation in alkalinity in the raw water, on the one hand, and the consumption of this alkalinity during nitrification in the biological tanks and the addition of $FeCl_3$ in the physicochemical stage, on the other hand, explain the fluctuation observed in the produced water. Compliance is 94% for this parameter.

5.2.3.5.4. Sulfates

The maximum chloride and sulfate concentrations observed in the effluent leaving the treatment plant are 85 mg.L^{-1} and 80 mg.L^{-1}, respectively. As the processes installed at the reuse plant do not treat sulfates, the same concentration is obtained in the produced water. Monthly analysis has shown that sulfate levels sometimes exceed the guarantee, which explains why compliance is 90%.

5.2.3.5.5. Iron

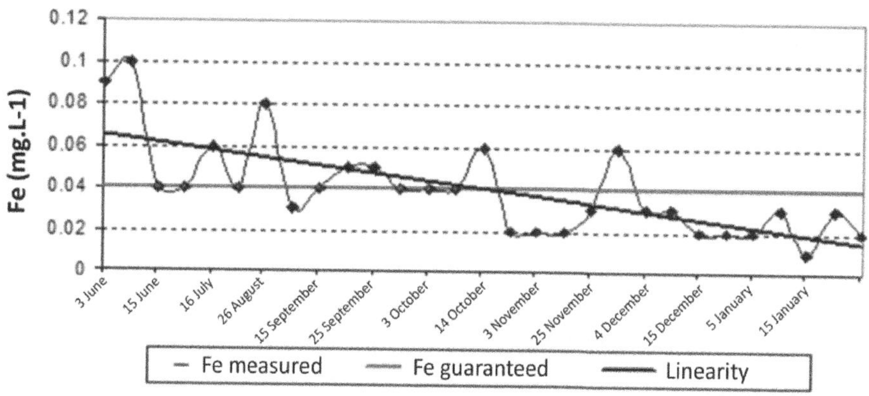

Figure 5.15. *Residual iron measured in produced water*

The concentration of iron in the effluent leaving the treatment plant is between 1.2 and 1.5 mg.L^{-1}. The quality of the water produced shows that the final iron concentration is between 0.03 and 0.045 mg.L^{-1}. Compliance for this parameter is 98%, which can be explained by ozone oxidation, which is very effective on this parameter.

5.2.3.5.6. Heavy metals

Compliance for all heavy metals is 100%. This result can be explained by the low variation of these elements in the raw water, their absorption by the bacteria present in the biological tanks and their precipitation in the lamella clarifier.

5.2.3.5.7. Forms of nitrogen

Compliance is 100% for all forms of nitrogen (ammonia, nitrate and nitrite). All are removed beforehand in the biological tanks where the ammonia and nitrate elimination sequences are carried out.

5.2.3.5.8. Bacteriology

Bacteriology covers total aerobic bacteria (TAB), total coliforms, molds, yeasts and worm eggs (Table 5.8).

Parameters	Unit	Guaranteed values	Average	Compliance (%)
Common germs	CFU/mL	100	14	100
Total coliforms	CFU/100 mL	0	0.1	98
Moulds	CFU/mL	0	0.1	90
Yeasts	CFU/mL	10	0.3	100
Helminth eggs	u/L	0	0	100

Table 5.8. *Bacteriological compliance of produced water*

Compliance for common germs, yeasts and helminth eggs is 100%. Figure 5.16 shows an average of 14 CFU/mL for common germs, with a maximum of 85 CFU/mL observed on two occasions.

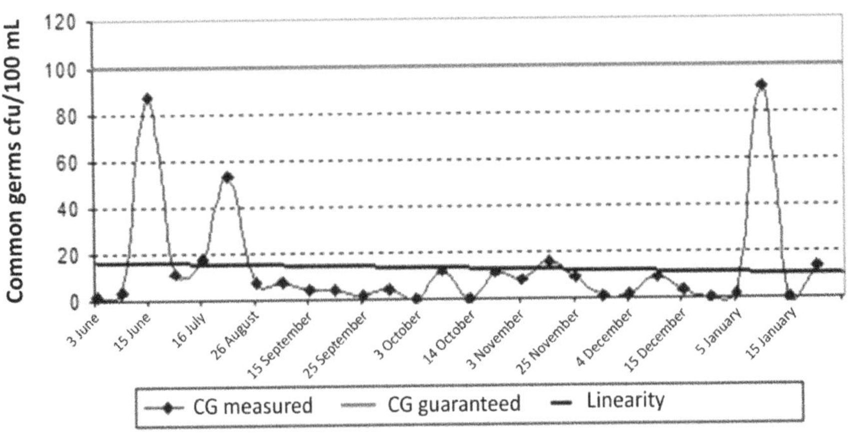

Figure 5.16. *Guaranteed and measured common germs*

Compliance for total coliforms was 98%, with an average of 0.1 CFU/100 mL.

The chlorine dosage was then increased to ensure total disinfection of coliforms and molds.

5.2.4. *Conclusion*

Drinking water that had previously been depleted by industry was made available and redistributed to previously disadvantaged suburban communities, without the need to invest in major bulk water supply and treatment infrastructures.

An additional 400,000 people were supplied. This sustainable wastewater management project with multiple environmental, economic and social impacts in South Africa has become an example of a solution that sees wastewater as an asset rather than a waste to be disposed of.

This wastewater reuse project has made a significant contribution to the preservation of natural water resources. The Durban project demonstrated that innovative approaches to water resource management, environmental management, wastewater treatment technology and institutional arrangements can deliver outstanding results. The project's success was based on innovation, thinking beyond traditional approaches and finding the winning formula for all participants.

It is also an example of a successful and innovative PPP to improve the sustainability of wastewater management, minimize environmental impact and deliver multiple benefits to the community. The city was able to turn a difficult situation into an opportunity, taking advantage of local conditions and innovative thinking that led to a win–win solution for all parties involved.

This has changed the behavior of industrialists, who no longer regard wastewater as a waste product, but as a potential resource, given the success achieved in Durban. In the South African context, the municipality has continued its "revolution" in its approach to the management and implementation of water projects.

Durban is South Africa's second most populated city, with over 1,100,000 inhabitants in a metropolis of almost 3.5 million. It is home to adherents of many faiths, as well as a wide variety of political, cultural, religious, ethnic and environmental groups who may have profound philosophical objections to wastewater reuse. There has also been fierce public concern where wastewater recycling has been considered, sometimes leading to initiatives being abandoned. The objectives of the project were to determine whether there were groups in

Durban with – specifically – religious or philosophical objections to the reuse of wastewater to make drinking water. The aim was to gain knowledge about the presence of objections likely to emerge from Durban's various communities. The findings of the surveys and discussions revealed that there appeared to be no fundamental religious objections to this use, and that local residents wanted the concept to be sustainable.

5.3. Kranji (Singapore): irrigation, indirect drinking water, specific industries

Singapore has been suffering from a water deficit for decades, and it has often relied on Malaysia to make up for this deficit since 1961. As early as 1998, the Public Utilities Board (PUB) and the Ministry of the Environment (ENV) launched specific studies to reuse wastewater as drinking water.

The first step was to determine the feasibility of using treated wastewater as a source of raw water to supplement Singapore's water supply. A pilot study at the Bedok site, spread over 2 years, tested a double-membrane process (microfiltration [MF]/reverse osmosis [RO]) followed by UV disinfection, for a flow rate of 10,000 $m^3.day^{-1}$. The feed water for the demonstration plant is clarified secondary effluent from an activated sludge treatment plant, with an average composition of 10 $mg.L^{-1}$ BOD5, 10 $mg.L^{-1}$ TSS, 6 $mg.L^{-1}$ ammonium nitrogen, 400–1,600 $mg.L^{-1}$ total dissolved solids (TDS) and 12 $mg.L^{-1}$ total organic carbon (TOC).

The treatment process includes micro-screening (0.3 mm), a 0.2 mm MF membrane filtration stage to remove suspended solids and fine particles, followed by a second RO membrane stage. The demineralized water is then disinfected with UV radiation (60 $mJ.cm^{-2}$).

Chlorine is added at two points before and after MF to control the rate of biological fouling of the membrane system. Dechlorination is carried out upstream of RO membranes.

Water recovery rates for CMF membranes (Evoqua, ex-Memcor) ranged from 85% to 90%, with an average of 87% (standard deviation ± 1.9).

The RO membrane workshop comprises two trains of membranes (three stages), handling a unit flow rate of 5,000 $m^3.day^{-1}$ configured for a recovery rate of 80–85%. The water produced is called NEWater.

As part of the sampling and monitoring program (SAMP), over 200,000 analyses were carried out, covering 296 physicochemical and microbiological parameters, 96 of which were referenced by the USEPA and 113 by the World Health Organization. Extensive toxicology tests on fish and mice were also carried out during the 2 years that the demonstration unit was in operation.

A final assessment was carried out by groups of international experts, who concluded that NEWater presented no health problems. This validation helped provide credibility and assurance to the authorities.

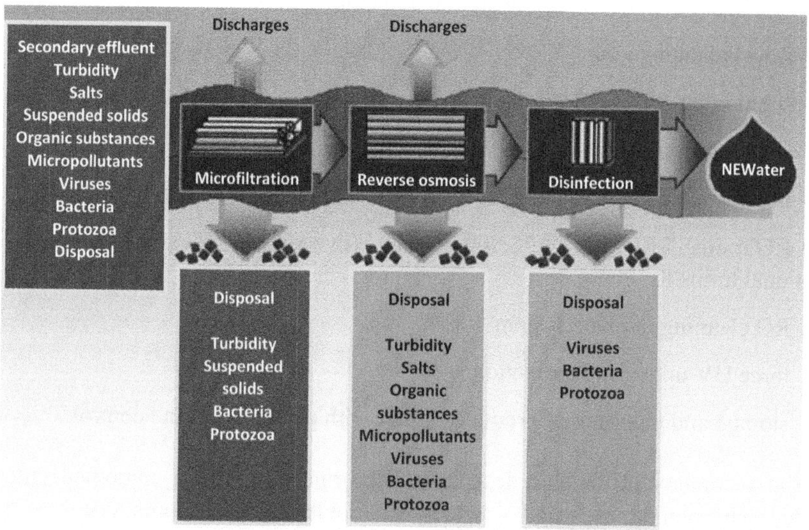

Figure 5.17. *Performance assessment of each stage of treatment*

In view of the satisfactory results, NEWater has agreed to mix treated wastewater with surface water to supply the new water reuse plant, which will deliver drinking water to Singapore on the premise of planned indirect drinking reuse.

In January 2002, PUB awarded Veolia the contract to install and commission Phase 1 (40,000 $m^3.day^{-1}$) of the new wastewater reuse plant, based on the same process as the Bedok demonstration plant.

The Kranji wastewater reuse plant reproduces the same process as the Bedok demonstration plant, albeit with a different type of MF membrane. CMF-S (submerged continuous MF) membranes from Evoqua (formerly Memcor) replace the pressurized membranes (CMF).

The plant was designed to allow capacity expansion up to a final Phase 2 capacity of 72,000 $m^3.day^{-1}$.

5.3.1. *Sizing the Kranji plant*

The Kranji plant comprises the following unit processes:

– secondary effluent pumping station and chlorine dosing;

– flow balancing tank;

– submerged MF (CMF-S);

– cleaning in place system;

– microfiltered water storage with additional chlorine dosing and dechlorination;

– RO trains: 5 RO trains of 8,000 $m^3.day^{-1}$ (two stages, 75% recovery), with four additional trains for Phase 2;

– RO cleaning in place system;

– three UV units for disinfection;

– storage and pumping of produced water with pH and chlorine control.

The secondary effluent is dosed with chlorine to create a combined chlorine residual (chloramine) of 3 $mg.L^{-1}$, which is variable due to ammonia levels.

The chloraminated secondary effluent is pumped from the balancing tank to pre-filters with a 500-μm cut-off, and then to the MF membrane workshop. A feed channel distributes the effluent to the submerged cells (CMF-S 480S10T) for Stage 1, with a further five cells planned for Stage 2. Filtration is performed in the external-internal direction.

The effluent is drawn through the 0.2-μm porous membranes by the filtrate pump. The pump rate is controlled to maintain a cell-specific filtrate flow rate that removes suspended particles, bacteria and protozoa from the effluent.

For Phase 1, the CMF-S system comprised a train of six CMF-S 480 S10T cells, each comprising 448 installed modules (14 racks, each containing 32 membrane sub-modules).

Each cell incorporates a filtration pump and a specific valve for individual flow control, backwash and CIP. The cell is drained by gravity into a wash sump.

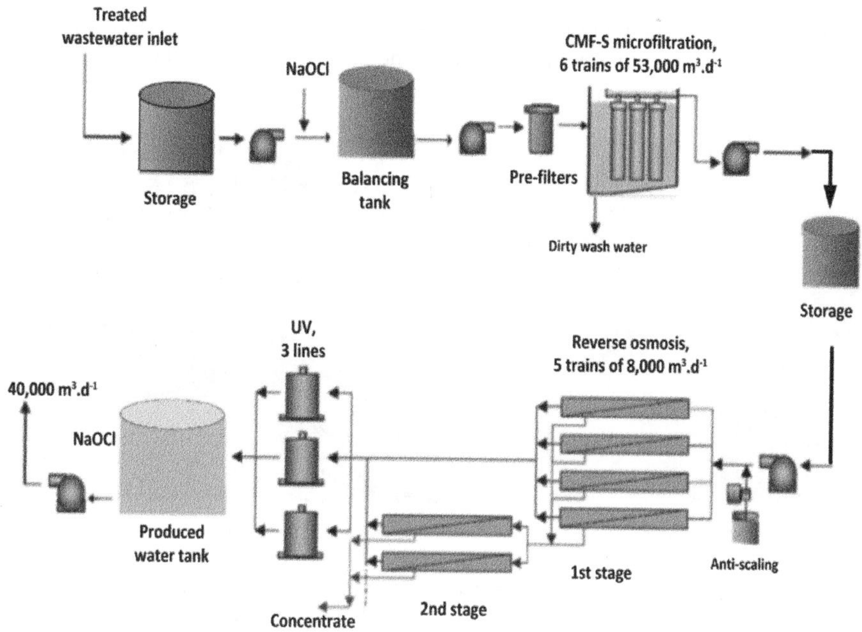

Figure 5.18. *Kranji wastewater reuse treatment plant (Singapore)*

Each CMF-S cell is backwashed after 20–45 min of filtration. Backwashing uses low-pressure air to clean and remove material accumulated on the fiber surface, while the wash water circulates through the membrane. The dirty water is discharged from the cell and recycled to the top of the Kranji treatment plant.

Every 3–6 weeks, the cells are chemically cleaned in situ using a clean-in-place (CIP) procedure that removes unextracted material by backwashing.

The filtrate produced in the CMF-S workshop flows to the downstream filtrate storage tank, then to the RO plant.

The RO membrane plant comprises five RO trains, each producing 8,000 $m^3.day^{-1}$. Each RO train consists of 2 stages operating at 75% recovery using Hydranautics LFC-1 elements. Each train comprises 49 pressure tubes of seven elements, while Stage 2 comprises 24 pressure tubes of seven elements. The RO workshop has its own CIP system. A sequestering agent is injected upstream of the RO membranes to prevent the precipitation of carbonates and other minerals.

The RO permeate is then directed to the UV disinfection stage, comprising three UV reactors. Two are in operation, while one is on stand-by. The UV dose is 90 mJ.cm^{-2}. The pH of the produced water is adjusted, before the storage tank.

a) b)

Figure 5.19. *CMF-S a) and reverse osmosis b) membranes*

5.3.2. *Performance*

5.3.2.1. *Raw water quality*

Table 5.9 shows the composition of effluent leaving the Kranji treatment plant.

Turbidity	NTU	12 (5–29)
Total chlorine dose	mg.L^{-1}	3 (1.5–5)
Conductivity	µS.cm^{-1}	538 (400–600)
Temperature	°C	30 (29–31)
Ammonia	mg.L^{-1}	3.5 (0.3–12)

Table 5.9. *Some parameters of the effluent leaving the Kranji treatment plant*

Table 5.10 shows the design parameters and the conditions and values observed at the plant.

CMF-S membranes performed well for over 25 days without the need for CIP, as shown in Figure 5.20.

Parameters	Sizing	Conditions observed on site	Remarks
Flow	<36 L.m^{-2}.h^{-1}	25.8–35.8 L.m^{-2}.h^{-1}	Variable flow depending on demand
Backwash frequency	25 min	35 min	May vary according to effluent quality
CIP frequency	>21 days	>21 days	–
Membrane integrity	4 LRV	4.4 LRV	–
Turbidity	<0.2 NTU 100% <0.1 NTU 95%	<0.2 NTU 100% <0.1 NTU 95%	Average 0.05 NTU
Particle count	2–5 µm.mL^{-1} 95%	–	Average 0.7 µm.mL^{-1}
SDI 15 min	<3	<1.5	Average 0.7
Recovery rate	90%	>92%	–

Table 5.10. *CMF-S workshop performance*

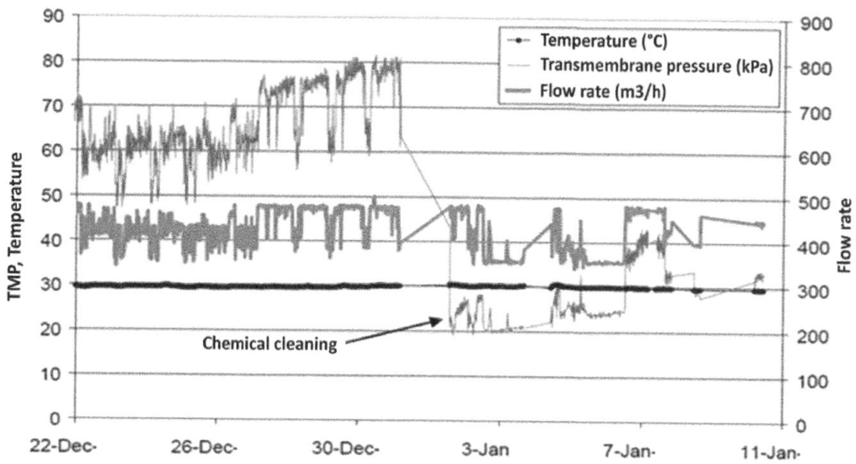

Figure 5.20. *Evolution of transmembrane pressure on CMF-S membranes*

5.3.2.2. *RO membrane workshop*

RO unit sizing is summarized in Table 5.11.

Parameters	Units	Bedok prototype	Kranji plant (Phase 1)
Number of trains	–	2	5
Number of pressure tubes	–	45 × 7	49 × 7
Unit area of a membrane	m^2	37.2	37.2
Recovery rate	%	75	75
Produced water flow	$m^3.day^{-1}$	10,000	40,000
Temperature	°C	30	30
Feed pressure	bar	8.2	7.5
Flow	$L.m^{-2}.h^{-1}$	17.6	26.1
Total dissolved salts feed	$mg.L^{-1}$	987	633
Permeate TDS	$mg.L^{-1}$	46.6	46.6

Table 5.11. *Sizing and performance of RO membrane workshop*

Table 5.12 shows the composition of the water produced by the RO membrane workshop.

Parameters	Guaranteed values	Measured values
Iron (Fe)	<0.05 $mg.L^{-1}$	<0.003 $mg.L^{-1}$
Boron (B)	<0.1 $mg.L^{-1}$	0.02 $mg.L^{-1}$
Manganese (Mn)	<0.05 $mg.L^{-1}$	<0.003 $mg.L^{-1}$
Fluorine (F)	<0.5 $mg.L^{-1}$	0.1 $mg.L^{-1}$
Nitrate (N)	<10 $mg.L^{-1}$	0.56 $mg.L^{-1}$
Ammoniacal nitrogen (N)	<0.5 $mg.L^{-1}$	<0.5 $mg.L^{-1}$
Conductivity	<150 µS/cm	22 µS/cm
pH	8.0–8.5	8–8.5
Silica (SiO_2)	<0.5 $mg.L^{-1}$	0.16 $mg.L^{-1}$
Total residual chlorine	0.5-1.0 $mg.L^{-1}$	<1 $mg.L^{-1}$
Aluminum	<0.12 $mg.L^{-1}$	<0.05 $mg.L^{-1}$
Chlorides	<5 $mg.L^{-1}$	<5 $mg.L^{-1}$
Sulfates	0.15–0.5 $mg.L^{-1}$	<0.2 $mg.L^{-1}$
Calcium	<0.5 $mg.L^{-1}$	<0.5 $mg.L^{-1}$
Turbidity	<0.2 NTU	<0.1 NTU
Total organic carbon	<0.1 $mg.L^{-1}$	0.08 $mg.L^{-1}$
Heterotrophic germs (at 35°C, 48 h)	<5 UFC/mL	<1 UFC/mL

Table 5.12. *Some parameters of produced water*

5.3.3. *Conclusion*

To ensure water security, Singapore has diversified its water resource alternatives within a long-term, forward-looking framework, enabling it to meet current and estimated future water needs. These strategies have been approved at the highest political level. Institutional and legal frameworks have been modified and improved in line with evolving needs.

NEWater is used for indirect drinking water purposes during dry periods by augmenting water sources in reservoirs. It mixes with raw water and undergoes conventional treatment before being distributed as tap water.

Figure 5.21. *Water output from UV disinfection unit*

The economic benefits of this application revolve around a growing industrial sector. Socially speaking, it is essential, as it provides water for domestic use during dry periods for the population, as well as supports jobs in the industrial and commercial sectors.

Currently, NEWater supplies around 40% of Singapore's water demand. It is expected to meet up to 55% of demand by 2060, by streamlining water infrastructure to collect 100% of wastewater.

Singapore introduced water reuse in 2003, at a time when Windhoek, Namibia and Orange County, California, were already producing reused water. Singapore studied their experiences, established its own system, and ensured widespread public acceptance through education and communication policies.

5.4. Illawarra (Australia): application, ocean protection, irrigation and industry

It is no surprise that Australia is a good case study for water recycling. Recognized as one of the driest continents on the planet, with one of the most variable climates, Australia offers its experience to other countries that are considering developing their wastewater management practices to maximize the benefits of this increasingly precious water source. Indeed, for over 40 years now, Australia has regarded wastewater as a resource rather than a waste product. The authorities insist on:

– the introduction of community programs to raise awareness of the benefits of recycled water;

– an increase in drinking water prices to enable reclaimed water to compete, rather than keeping prices artificially low;

– research programs covering the public health, environmental and social aspects of recycled water, so as to build and generate community trust.

Four decades later, the creation and adoption of national water recycling guidelines and a significant increase in recycled water references have backed this strategy. Today, problems of climate change and population growth confirm that it is more important than ever to persevere along this path:

– Australia's population has increased by 50%, mainly in the large coastal conurbations.

– Australia suffered its worst-ever drought.

– There have been significant changes in community awareness and attitude towards water recycling. Surveys of Australians during the 2000s have consistently shown that there is very strong community support for increasing the use of recycled water for countless non-drinking purposes.

The onset of the great Millennium Drought in Australia has brought the value of recycled water into sharper focus. Rather than seeing wastewater as waste, under pressure from communities, water utilities began to see it as a valuable source of water. The change in attitude was aided by the Council of Australian Governments (COAG), which supported an intergovernmental agreement on the National Water Initiative, and encouraged wastewater reuse and recycling where it was cost-effective.

Despite the cost of building reused water pipe networks and pumping water, the number of reused water plants has significantly increased. The increase in recycled water volumes was in line with climate requirements, and it was strongly supported by the community.

A major wastewater treatment and reuse project was implemented in Wollongong. It defined clear objectives as follows:

– improve water quality on Illawarra's beaches, particularly those located near wastewater treatment plants;

– protect coastal waters and reduce the impact on sensitive marine ecosystems, and minimize the negative impacts of effluents when discharged into the environment;

– provide high-quality recycled water for industrial reuse;

– save drinking water.

The project consisted of:

– modernization and expansion of Wollongong STP by installing a complete tertiary treatment, including UV disinfection;

– conversion of the existing Bellambi and Port Kembla wastewater treatment plants into rainwater treatment plants to store and treat wastewater;

– the construction of a wastewater reuse plant using a membrane process, including MF and RO technologies to recycle 20,000 $m^3.day^{-1}$ for the Blue Scope Steel industry in Port Kembla;

– construction of wastewater pumping stations and transfer pipes to transport flows from the Bellambi and Port Kembla wastewater treatment plants;

– construction of a new outfall to the ocean with two 1-km pipes at Wollongong.

Veolia was commissioned by Sydney Water Corporation to design, build and provide operational advice for the upgrading of three wastewater treatment plants in the south of Sydney (Bellambi, Port Kembla, Wollongong) to serve 300,000 people.

This is one of the largest wastewater reuse facilities in Australia, with the dual environmental benefits of reducing the volume of ocean wastewater discharged by 40%, and reducing the use of drinking water in the steelmaking process by over 60%.

5.4.1. *Characteristics of wastewater treatment plants*

The Wollongong plant had to have the capacity to treat wastewater under several weather conditions. The flow rates of the connected plants and the catchments are summarized in Table 5.13.

Average wastewater characteristics are summarized in Table 5.14.

Sites	Flow in $m^3.j^{-1}$		
	Average dry weather	Maximum dry weather	Rainy weather
Bellambi transfer to Wollongong	23,000	41,000	69,000
Port Kembla transfer to Wollongong	17,000	38,000	51,000
Wollongong catchment	19,000	43,000	200,000
Total: Wollongong increase	59,000	122,000	320,000

Table 5.13. *Flow connected to the Wollongong (Illawarra) treatment plant*

Parameters	Unit	Average concentration	Maximum concentration
Temperature	°C	16	24
BOD5	$mg.L^{-1}$	250	325
COD	$mg.L^{-1}$	550	715
TSS	$mg.L^{-1}$	250	325
Fats	$mg.L^{-1}$	50	–
Ammonia	$mg.L^{-1}$	45	–
Total Kjeldahl nitrogen	$mg.L^{-1}$	55	71.5
Total phosphorus	$mg.L^{-1}$	10	13
Alkalinity ($CaCO_3$)	$mg.L^{-1}$	220	–
Fecal coliforms	CFU/100 mL	10^8	–

Table 5.14. *Average raw wastewater characteristics*

5.4.2. *Treatment processes*

After primary treatment, wastewater from the three plants is collected at the new Wollongong wastewater treatment plant.

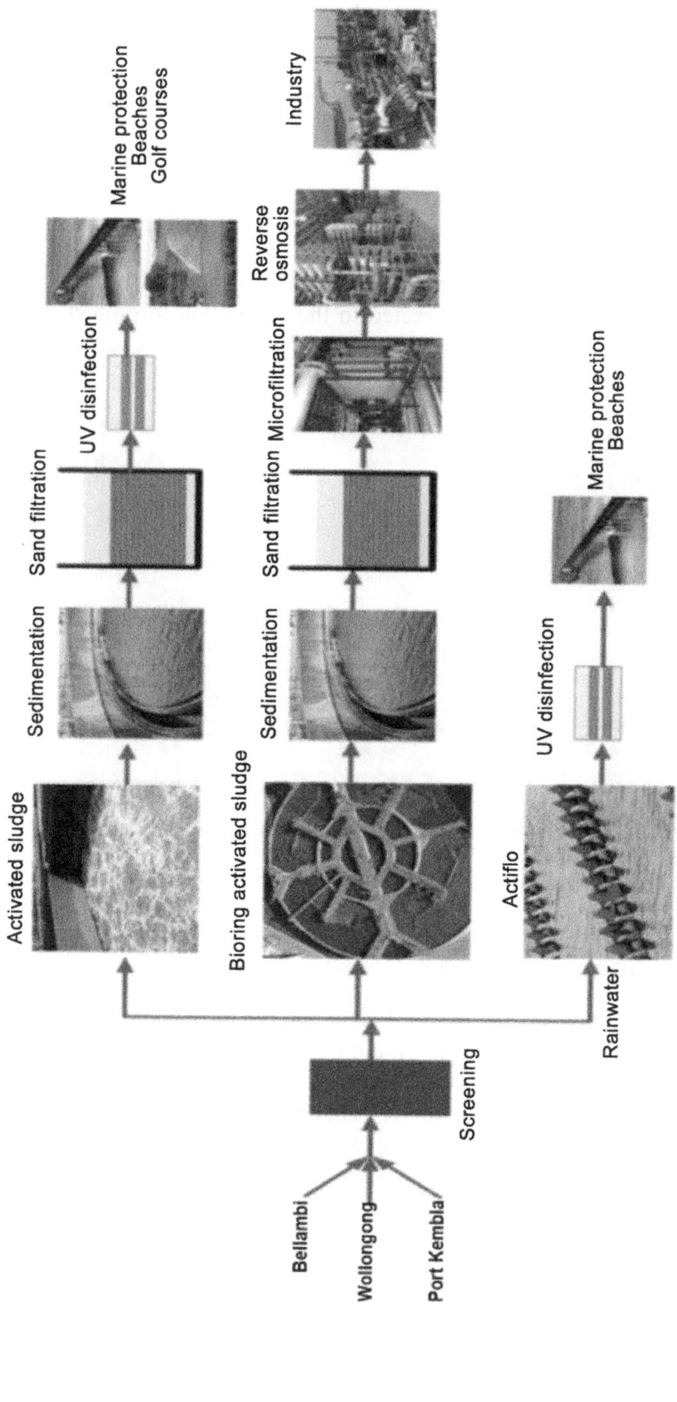

Figure 5.22. *Wollongong's three treatment processes*

The effluent leaving the main collector, which collects flows from the three plants, undergoes physical screening and is then divided into three flows corresponding to the three uses, as shown in Figure 5.22:

– The first flow is treated by an activated sludge process, followed by sedimentation, sand filtration and UV disinfection. It is used for golf course irrigation or discharged into the ocean.

– The second flow is treated by an activated sludge process, followed by sedimentation, sand filtration, MF and RO. It is reused for industrial purposes.

– The third flow (rainwater) is treated by a physicochemical process (Actiflo), followed by UV disinfection, then directed to the sea outfall to protect the marine environment and beaches. This process receives all flows in excess of 177,000 $m^3.day^{-1}$. It guarantees effluent treated to 15 $mg.L^{-1}$ TSS.

Figure 5.23. *General view of the Wollongong treatment plant (Illawarra, Australia)*

5.4.2.1. *High-performance process*

The membrane process is the most efficient and the most complex. It is designed to produce water of the highest quality. It comprises biological treatment designed to remove most organic pollution (BOD5, COD, nitrogen and phosphorus). Sedimentation removes suspended solids. Sand filtration (12 $m.h^{-1}$ in N^{-1}) refines turbidity. Filtered water quality is as follows: TSS: 5 $mg.L^{-1}$ (353 days/year) to 20 $mg.L^{-1}$ (12 days/year); BOD5: 5.5 $mg.L^{-1}$; total COD: 36 $mg.L^{-1}$; temperature: 16–24°C.

This is followed by the MF and RO stages, which deliver water that meets the standards required by industry (Table 5.15).

5.4.2.1.1. MF workshop

The plant is designed with 8 MF blocks in operation and one on standby. Each block has 108 membrane modules and is designed to produce a maximum of 26,800 $m^3.day^{-1}$:

– nine E112M10C units;

– membrane surface area of nine units: 15,120 m^2;

– net flow rate produced: 23,530 $m^3.day^{-1}$;

– washing frequency: 22 min (duration: 3 min);

– CIP frequency: caustic soda: >7 days; acid: >21 days;

– instantaneous flow rate: 1,167 m^3/h;

– instantaneous flux: 73.85 $L.h^{-1}.em^{-1}$ (1 em = 1 m^2 nominal membrane) at 16°C. At 20°C: 96.4 $L.h^{-1}.em^{-1}$;

– recovery rate: 85%.

The quality of the filtrate is independent of the CMF feed water quality and offers the following guarantees:

– turbidity < 0.1 NTU;

– SDI < 3;

– LRV of 4 log for particles > 0.2 μm;

– total coliforms < 1 CFU/100 mL;

– fecal coliforms < 1 CFU/100 mL;

– parasites: 4 log LRV;

– viruses: 3 log LRV.

5.4.2.1.2. Reverse osmosis

The RO workshop comprises five trains, each producing 5,000 $m^3.day^{-1}$, in other words, 125% of the initial capacity. Each train consists of three stages and offers a recovery rate of 85%.

Full production capacity is 20,000 $m^3.day^{-1}$ with one train on stand-by (cleaning or maintenance). This guarantees the production of reused water for the industrial sector.

a) b)

Figure 5.24. *Microfiltration workshop a) and reverse osmosis workshop b)*

Table 5.15 shows the quality of effluent produced by the water reuse plant for local industry.

Parameters	Unit	Discharge to ocean	To reuse plant	Rainwater	Compliance (%ile)
BOD5	$mg.L^{-1}$	8	10	–	90
TSS	$mg.L^{-1}$	8	10	–	90
Ammonia	$mg.L^{-1}$	–	1	–	50
Total phosphorus	$mg.L^{-1}$	1	1	–	50
Enterococci	CFU/100 mL	35	35	10^4	50
Fecal coliforms	CFU/100 mL	150	150	10^4	50

Table 5.15. *Water quality required at treatment plant outlets*

a) b)

Figure 5.25. *Port Kembla a) and Wollongong outfall b)*

5.4.2.2. *Water quality requirements for reuse*

Parameters	Units	Values	Parameters	Units	Values
Aluminum	$mg.L^{-1}$	<0.1	**Sodium**	$mg.L^{-1}$	<15.3
Ammonia	$mg.L^{-1}$	<1	**Strontium**	$mg.L^{-1}$	<0.05
BOD5	$mg.L^{-1}$	–	**Sulfate**	$mg.L^{-1}$	<1
Barium	$mg.L^{-1}$	<0.05	**TSS**	$mg.L^{-1}$	<1
Calcium	$mg.L^{-1}$	<10	**Total dissolved salts**	$mg.L^{-1}$	<50
Chloride	$mg.L^{-1}$	<15	**Total phosphorus**	$mg.L^{-1}$	<0.15
Fluorine	$mg.L^{-1}$	<0.15	**Zinc**	$mg.L^{-1}$	<0.1
Iron	$mg.L^{-1}$	<0.05	**Hardness ($CaCO_3$)**	$mg.L^{-1}$	>15–<20
Magnesium	$mg.L^{-1}$	<3	**Total coliforms**	CFU/100 mL	<1
Manganese	$mg.L^{-1}$	<3	**Fecal coliforms**	CFU/100 mL	<1
Nitrate	$mg.L^{-1}$	<1.3	**Viruses**	u/50 L	<1
Nitrite	$mg.L^{-1}$	<0.05	**Protozoa**	u/50 L	<1
pH	–	7–8.5			
Potassium	$mg.L^{-1}$	<2			
Silica	$mg.L^{-1}$	<0.5			

Table 5.16. *Physicochemical and microbiological characteristics of water to be reused*

5.4.3. *Conclusion*

The Wollongong wastewater treatment and reuse plant is the heart of Illawarra's wastewater infrastructure. It is a complex plant, designed to meet stringent quality requirements for different uses. Compact treatment processes were chosen over more conventional treatment processes because of the limitations imposed by the footprint. With a treatment plant including tertiary treatment, an independent storm water treatment line and a high-performance process including MF and RO membranes for specific use, Wollongong is the most advanced treatment in the southern hemisphere. The project has restored the picturesque Bellambi headland, making it available for public use for recreational purposes. Wollongong brings environmental protection to the magnificent Illawarra coastline, protecting coastal areas, reducing the impact on sensitive marine ecosystems and minimizing the negative impacts of effluent that is discharged into the environment. In addition, the plant provides a water supply for local industry, saving precious drinking water. Tedious and compulsory water restrictions, due to chronic drought, are part of everyday life. Wollongong's water savings are a benefit to the local population.

As the pressures of population growth, urbanization and climate change continue, the demand for water recycling will increase in the future. Wollongong provides Australian authorities with excellent operational data to design new systems that protect the environment, reduce the water industry's carbon footprint and are acceptable to our communities.

In the future, urban wastewater reuse systems:

– will consume less energy and contribute much less to GHG emissions;

– be better integrated into urban planning;

– increase drinking water quotas for populations by reserving a large proportion of reused water for industry.

There will be a growing trend to recover other valuable by-products from wastewater, such as phosphorus, nitrogen, potassium and other basic chemicals.

5.5. Honolulu: irrigation and industrial applications

Surrounded by an ocean, the Hawaiian Islands are limited in their natural freshwater resources. The main sources of readily exploitable drinking water are groundwater springs, which are used for both domestic purposes and sugarcane irrigation, although irrigation water does not require as high a quality as drinking water.

The island of Oahu's growing global freshwater needs were expected to exceed the potential yield of underground freshwater sources, according to Honolulu Board of Water Supply projections made in 2000. It was recognized that reusing wastewater effluent for irrigation would increase natural water resources, provide additional or alternative fertilizers and reduce pollution of ocean waters.

The City and County of Honolulu contracted Veolia in 1998 to build and operate the Honouliuli Water Reclamation Facility, which is designed to treat 59,000 m^3 of wastewater to produce 54,500 m^3 of reused water.

The treatment process includes:

– a biological wastewater effluent treatment stage;

– sand filtration;

– a pressurized MF membrane workshop;

– an RO membrane workshop;

– UV disinfection;

– final chlorination.

The process generates two qualities of water: high-quality water produced by RO, which is sold to industrialists for electricity and oil refining in the Campbell Industrial Park, and water from the MF workshop, which is used for irrigation of golf courses and communal areas near the booming town of Kapolei.

Figure 5.26. *Reverse osmosis membrane workshop*

The water qualities produced are summarized in Table 5.17.

This plant has saved 45,360 m^3 of drinking water per day. Industries using this water have released fresh water to underserved communities. In addition, this strategy improves water security during periods of drought.

– For the city, it creates a long-term source of revenue from a pre-levied tax on water production and reduces the city's operating costs.

– For manufacturers, it provides a complementary solution to meet their requirements at a competitive price. The actual price is lower than that of drinking water.

– For the environment, it ensures better protection in this popular and exotic vacation spot and tourist residence. It also reduces pollution from wastewater discharged into the sea.

Microfiltration		Reverse osmosis	
Turbidity	<1 NTU	Silica	<2 mg.L^{-1}
TSS	<2 mg.L^{-1}	TSS	<2 mg.L^{-1}
pH	6–7	pH	6–7
Total dissolved salts	<500 mg.L^{-1}	Total dissolved salts	38 mg.L^{-1}
Chlorides	<1,250 mg.L^{-1}	CO_2	5 mg.L^{-1}

Table 5.17. *Water quality after microfiltration and reverse osmosis*

This facility is the largest wastewater reuse plant in the Hawaiian Islands and uses state-of-the-art technology to treat secondary effluent, which was previously discharged into the Pacific Ocean.

Figure 5.27. *General view of the Honolulu wastewater treatment and reuse plant*

By dedicating this non-drinking supply to industries, the community preserves its limited drinking water resources through rigorous treatment and reuse of wastewater, generating a lasting benefit for taxpayers, the city, the environment and a number of businesses, which are served by a quality water supply.

This plant helps to protect precious water resources and demonstrates how creative thinking and resource management can be a win–win situation.

In Honolulu, the circular economy is based on waste control and the reuse and recycling of everything produced. The wastewater treatment plant is essential, enabling the surrounding community to develop without running out of drinking water and to withstand potential shortages caused by climate change.

In Hawaii, wastewater reuse has demonstrated multiple benefits for irrigation, as well as for reducing demand on drinking water aquifer sources. The Honolulu Board of Water Supply (HBWS) has taken a cautious approach to authorizing the use of recycled water on drinking water aquifers due to concerns about groundwater degradation or contamination. The HBWS was concerned about the migration of organic constituents such as pesticides and other micropollutants. As a result, the HBWS developed a project to study the potential impacts of irrigation with recycled water in Central Oahu.

5.6. Gerringong and Gerroa (Australia): controlled irrigation and beach protection applications

Gerringong and Gerroa are coastal towns of 3,500 inhabitants, located 120 km from Sydney on Australia's southeast coast (New South Wales). The region is a popular vacation destination, well known for its diverse flora and fauna, as well as its beaches. The main local industries are dairy farming and tourism. A public request has been made to minimize the negative impacts of effluent when discharged into the environment, especially as prior to the start of the project in June 2001, the area had no reticulated system and all wastewater was discharged into septic tanks. Poor maintenance of these systems led to a rapid deterioration in effluent quality. This delicate situation entailed potential health risks and contributed to the degradation of water quality in local streams and bathing waters. Facing this, the Gerringong-Gerroa sanitation system was implemented in 2001, consisting of the construction of a wastewater treatment plant in the Gerringong-Gerroa area, which became operational in 2002. It was designed to meet the needs of the local community, estimated at 11,000 inhabitants.

The treatment and reuse plant is located near the Crooked River, and is contractually managed by Sydney Water.

Figure 5.28. *Location of the towns of Gerringong and Gerroa*

5.6.1. *Treatment process and guarantees*

The plant is equipped with secondary and tertiary treatment:

– primary treatment: screening, sand removal, de-oiling;

– secondary treatment: biological treatment, clarification;

– tertiary treatment:

  - sand filtration (number 5),
  - ozonation, four GAC filters,
  - MF,
  - medium-pressure UV disinfection (2),
  - final chlorination.

The quality of the water produced is equivalent to that of the drinking water distributed.

The final effluent is stored in a 50,000 m^3 reservoir before being pumped to irrigate 120 ha of pasture, feeding a dairy farm adjacent to it.

When irrigation is not possible and the storage tank is full, high-quality effluent is discharged into a dune or wetland area.

The overall layout of the treatment process is shown in Figure 5.29.

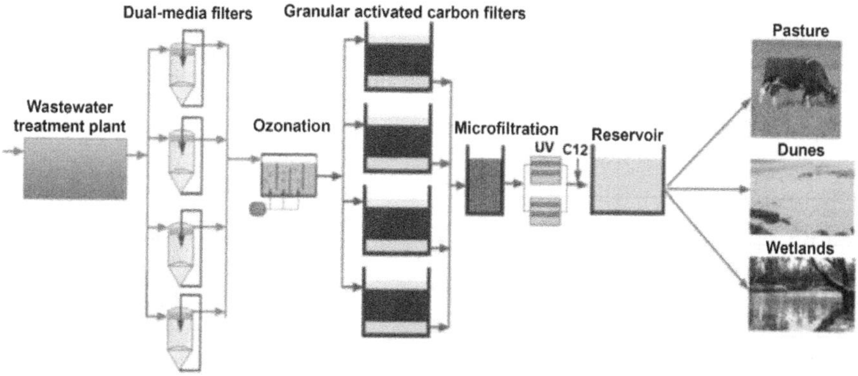

Figure 5.29. *General diagram of the Gerringong-Gerroa wastewater treatment system*

5.6.2. *Microfiltration workshop*

Raw effluent parameters	
BOD5	280 $mg.L^{-1}$
TSS	295 $mg.L^{-1}$
Total N	55 $mg\ N.L^{-1}$
Total P	10 $mg\ P.L^{-1}$
COD	644 $mg.L^{-1}$
$N-NH_3$	45 $mg.L^{-1}$
Alkalinity	280 $mg\ CaCO_3.L^{-1}$

Table 5.18. *Physicochemical characteristics of raw effluent*

The sizing of the MF unit is based on the following elements:

– variable flow rate between 28 and 34 $L.s^{-1}$;

– number of modules: 108; material: polypropylene; cut-off: 0.2 µm;

– unit area per module: 33.8 m^2;

– average flow: 27.6 $L.h^{-1}.m^{-2}$;

– backwash frequency: every 20 min;

– CIP triggered when TMP reaches 100 kPa, or every 6 months.

The final guarantees on disinfected water are shown in Table 5.19.

	Contractual guarantees	**Values obtained**
BOD5P	5 $mg.L^{-1}$	<2 $mg.L^{-1}$
TSS	<1 $mg.L^{-1}$	<2 $mg.L^{-1}$
Total N	3 mg $N.L^{-1}$	0.87 mg $N.L^{-1}$
Total P	0.1 mg $P.L^{-1}$	0.04 mg $P.L^{-1}$
pH	6.5–8	6.6–8.3
Fecal coliforms	<1 CFU/100 mL	0

Table 5.19. *Guaranteed quality of produced water*

The plant also produces Class A biosolids, which are recycled and used for land application.

a) b)

Figure 5.30. *Continuously washed sand filter a) and microfiltration b)*

5.6.3. *Conclusion*

As development and tourism in the area have increased, pollution of local waterways (Crooked River and Blue Angle Creek), lagoons (Werri Lagoon) and beaches has been significantly reduced, having a positive impact on the region. In addition, the advanced wastewater treatment system is regarded as one of Australia's most innovative environmentally sustainable wastewater systems, using best practice water treatment processes.

The advanced wastewater treatment scheme enables water to be reused on neighboring agricultural properties.

5.7. Playgrounds and leisure parks (France)

Playgrounds and leisure parks welcome millions of visitors every year. They use around 2,000,000–2,500,000 m^3 of water per year as follows:

– supply water for water games and ponds;

– watering green spaces and golf courses;

– supply water to restaurants and hotels;

– washing roads and pedestrian walkways;

– supply various technical circuits (heating and cooling water).

Parks have their own wastewater treatment plant and reuse treated water. In view of the likely development of algae in the water basins, a P requirement of less than 0.05 mg.L^{-1} was requested. The physicochemical way, with $FeCl_3$ injection and gravity or sand filtration, was an option. However, the presence of other elements such as chlorides and sulfates led to a solution involving RO membranes.

The various processes include:

– raw water storage to regulate flow;

– an MBR with activated sludge and submerged MF membrane;

– treated water storage for membrane cleaning;

– an RO workshop for refining the quality of produced water;

– UV disinfection;

– chlorine make-up, if required;

– produced water storage.

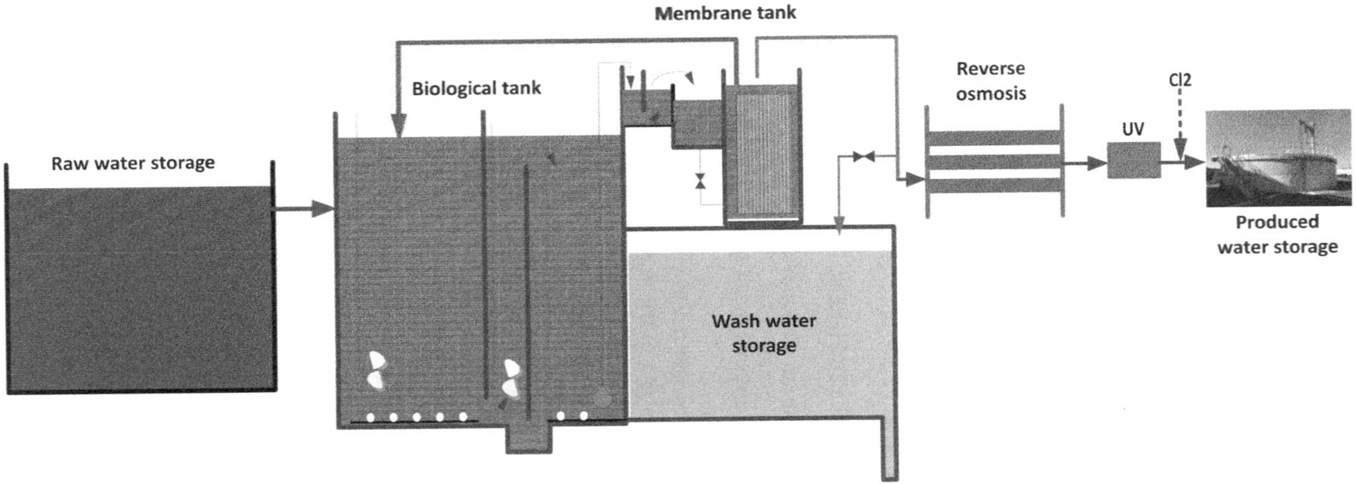

Figure 5.31. *Diagram of the wastewater reuse plant*

The sizing of the membrane workshop is based on the following elements:

– Membrane type: Zenon ZeeWeed PVDF:

- number of membrane lines: 3;
- number of modules per train: 96;
- membrane area per module: 31.6 m^2;
- membrane area per tank: 3,033.6 m^2;
- total membrane surface area of the plant: 9,100.8 m^2.

– Process data:

- maximum permissible flow rate in MBR workshop: 210 $m^3.h^{-1}$;
- air flow rate for membrane aeration: 2,700 $Nm^3.h^{-1}$;
- recirculation rate: 500%;
- minimum filtration flow rate: 20 $m^3.h^{-1}$;
- maximum filtration flow rate: 70 $m^3.h^{-1}$;
- backwash flow rate: 100 $m^3.h^{-1}$.

Table 5.20 shows raw and produced water qualities.

Parameters	Unit	Average raw water	Average MBR treated water	Average produced water
COD	$mg.L^{-1}$	448	26.5	<5
BOD5	$mg.L^{-1}$	198	≤5	<2
Total organic carbon	$mg.L^{-1}$	–	≤5	<1
TSS	$mg.L^{-1}$	224	<3	–
Total N	$mg.L^{-1}$	42.6	7.8	<2
N-NH_4	$mg.L^{-1}$	33.1	0.6	<0.5
N-NO_2	$mg.L^{-1}$	0.1	0.0	0
N-NO_3	$mg.L^{-1}$	0.2	6	<0.1
Total P	$mg.L^{-1}$	9.1	0.3	0.02
Total coliforms	CFU/100 mL	10^5–10^6	<10	0
Fecal coliforms	CFU/100 mL	10^4–10^5	<10	0
Enterococci	CFU/100 mL	10^3–10^4	<10	0
Legionella	CFU/100 mL	10^3–10^4	<100	0
Protozoa	u/100 mL	10^2–10^3	<10	0

Table 5.20. *Composition of raw wastewater, and produced and distributed water*

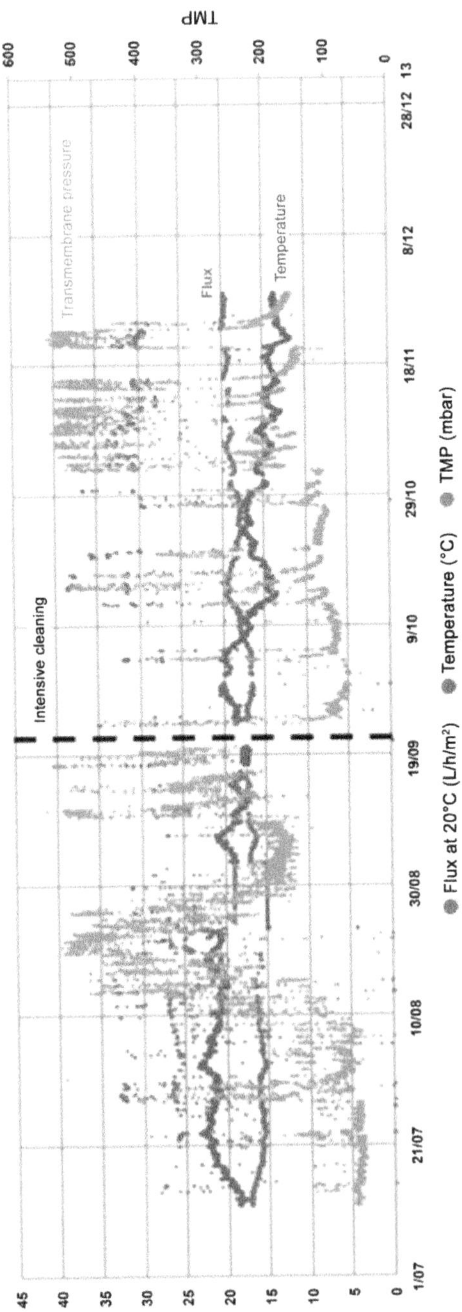

Figure 5.32. *(A) A year of MBR operations*

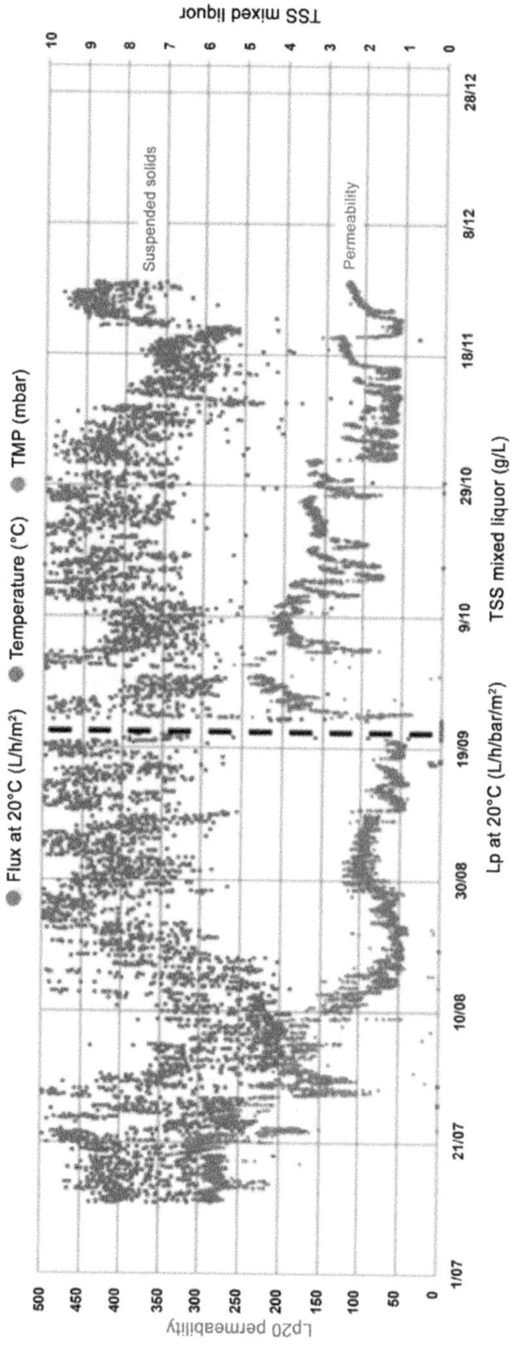

Figure 5.32. *(B) A year of MBR operations (continued)*

Figure 5.33. *Two applications for reused water*

5.8. References

Abi-Samra, B.C., Cook, P., Lange, P., Zepeda, J. (2002). Design and operation of a microfiltration and reverse osmosis system treating secondary effluent and providing high quality reclaimed water. In *Proceedings of the AMTA Conference*. Antenna Measurement Techniques Association, Tampa, FL.

Alexander, K., Alt, S., Owens, E., Patel, M., McGovern, L. (2003). Low fouling reverse osmosis membranes: Evidence to the contrary on microfiltered secondary effluent. In *Proceedings of the AWWA Membrane Technology Conference*. Antenna Measurement Techniques Association, Tampa, FL.

Bartels, C., Wilf, M., Andes, K., Iong, J. (2005). Design considerations for wastewater treatment by reverse osmosis. *Water Science & Technology*, 51(6–7), 473–482.

Cazurra, T. (2008). Water reuse of South Barcelona's wastewater reclamation plant. *Desalination*, 218, 43–51.

Esteban, R.I. and Ortega de Miguel, E. (2008). Present and future of wastewater reuse in Spain. *Desalination*, 218, 105–119.

Franks, R., Papukchiev, U., Murkute, P. (2004). MF/RO membrane system for treatment of secondary effluent wastewater. In *Proceedings of the Annual Water Reuse Conference*. Annual Water Reuse & Desalination, Phoenix, AZ.

Friedrich, E., Pillay, S., Buckley, C.A. (2009). Environmental life cycle assessments for water treatment processes: A South Africa case study of an urban water cycle. *Water S.A.*, 35(1), 73–84.

Gaid, K. (2006). Comparison of various, technologies for full scale reclamation water plants. In *9th Conference of International Water Assoc*. Intern. Desalination Assoc., Windhoek.

Gaid, K. (2012a). La réutilisation des eaux usées : quelles solutions pour quels objectifs ? *L'eau, l'industrie, les nuisances*, 299, 39–44.

Gaid, K. (2012b). Wastewater reuse: What solutions for what objectives. In *IWA Regional Conference on Wastewater Purification and Reuse*. Inter. Water Assoc., Heraklion.

Gaid, K. and Nugier, L. (2009). Traitement tertiaire des eaux usées en France : le cas de l'usine de l'Hermitage (La Réunion). *L'eau, l'industrie, les nuisances*, 322, 37–42.

Gaid, K., Bonne, P., Bacquet, G., Schrotter, J.C. (2004). Les technologies membranaires dans la réutilisation des eaux usées. *L'eau, l'industrie, les nuisances*, 262, 45–50.

Gaid, K., Durham, B., Etienne, S., Luck., F. (2005). Le recyclage de l'eau et la gestion intégrée du petit cycle de l'eau. In *Congrès de l'ASTEE*. ASTEE., Paris.

Grobicki, A. and Cohen, B. (1998). Water reclamation for direct reuse in urban and industrial applications in South Africa, and its projected impact upon, water demand. In *Water Research Commission, TR (KV118/99)*. WRC, Gezina.

Haarhoff, J. and Van der Merwe, B. (2009). Twenty-five years of wastewater reclamation in Windhoek, Namibia. *Water Science and Technology*, 33, 25–35.

Jacquemet, V. and Koning, E. (2006). Organic matter evolution in the treatment process of the New Goreangab Water reclamation plant at Windhoek, Namibia, Anjou. Report, Veolia Water, Paris.

Kimura, K., Amy, G., Dreses, J., Heberer, T., Kim, T.U., Watanabe, Y. (2003). Rejection of organic micropollutants disinfection by-products, endocrine disrupting compounds. *Journal of Membrane Science*, 227, 113–121.

Kivit, C. (2004). The study of organic carbon removal at the New Goreangab water reclamation plant. Master's Thesis, Delft University of Technology, Delft.

Kück, K., Freeman, M., Gisclon, A., Marche, O. (2002). Water reclamation project in environmental stressed area. In *Paper Presented at the Biennial Conference of the Water Institute of Southern Africa (WISA)*, Durban.

du Pisani, P. (2006). Direct reclamation of potable water at Windhoeck's Goreangab reclamation plant. *Desalination*, 188, 79–88.

du Pisani, P. and Menge, J. (2013). Direct potable reclamation in Windhoek: A critical review of the design philosophy of new Goreangab drinking water reclamation plant. *Water Supply*, 13(2), 214–226.

Salgot, M. (2008). Water reclamation, recycling and reuse: Implementation issues. *Desalination*, 218, 190–197.

Seah, H., Poon, J., Leslie, G., Law, J.B. (2003). Singapore's NEWater demonstration plant. *Water*, June, 43–46.

Singapore Public Utilities Board (2002). Singapore water reclamation study: Expert panel review and findings. Report, Singapore's National Water Supply Agency, 55.

Thompson, M. and Powell, D. (2003). Case study – Kranji high grade water reclamation plant, Singapore. Report, IMSTEC, Sydney.

Wilas, J., Draszawka-Bołzan, B., Cyraniak, E. (2016). Wastewater reuse. *World News of Natural Sciences*, 5, 33–41.

6

What Does the Future Hold for Wastewater Reuse?

6.1. Challenges and prospects

Water scarcity is increasing under the pressure of population growth, climate change and increasing pollution. Currently, one in nine people lacks access to improved sources of drinking water, and one in three lacks access to improved sources of sanitation. According to the 2019 Global Risks report provided by the World Economic Forum, more than 650 million people in 500 cities are expected to face a decline of over 10% in freshwater availability by the 2050s, particularly in the Middle East and North Africa. However, around 80% of all wastewater is discharged into the world's rivers. Under these conditions, water reuse can be seen as a promising approach to not only purify wastewater, but also to meet populations needs.

The total volume of reused treated wastewater worldwide is around 18 billion $m^3.year^{-1}$, as reported in *Global Water Market*, which represents 5–7% of the total volume of domestic wastewater (250–350 billion $m^3.year^{-1}$). However, the proportion of reuse differs considerably between different countries depending on local conditions. In areas of extreme water scarcity, such as the Middle East, 23% of collected wastewater is reused, mainly for irrigation. The government of Singapore, with its determination to be free of water demand, achieves 40% reuse of total wastewater. Countries such as Israel and Tunisia remain in the lead, with wastewater reuse rates exceeding 80%. Furthermore, the United States holds more than one-fifth of the world's reused water capacity, even though it only provides a national water reuse rate of 9.7%. In Europe, around 2.4% of all treated urban effluent is reused.

In this global situation, the wastewater reuse approach is a market that has a high potential for growth. As the European Parliament estimated in its May 2020 proposal, increasing wastewater reuse in agricultural irrigation from 1.7 billion $m^3.year^{-1}$ to 6.6 billion $m^3.year^{-1}$ by 2025 could reduce water stress by 5%. Implementing wastewater reuse requires substantial investments that many countries are unable to meet. All the more so since their first priority is to treat wastewater in a wastewater treatment plant and reduce discharges into the environment. This is their current priority.

In the global wastewater reuse market, over 90% of water production is reused for non-drinking purposes, including 52% for irrigation and 20% for industrial process water recirculation. In contrast, only 2.3% of water production is used for drinking purposes, and most of this for indirect reuse as drinking water. For example, the NEWater in Singapore's drinking reuse application cannot be used directly, but must be mixed with surface water in tanks before it can be considered for drinking use.

Significant progress has been made in the development of reliable technologies and analytical methods for monitoring the quality of reused wastewater. Continued research and demonstration efforts will result in further progress in the development of water reuse applications.

Despite this, the distribution of reused drinking water is low (2.3%), and this is mainly attributed to public concerns about health risks and the high cost of advanced treatment processes. Scientists play an important role in winning public trust.

The main concerns that the public have with regard to the assessment of health risks is the traces of contaminants in reused treated wastewater, as well as the means of monitoring to assess microbiological quality. It has been shown that drug molecules are mostly found in the soil, leaves and roots of plants that have been irrigated with treated wastewater, with fruits and tubers being less contaminated. This study concludes that it would be worthwhile to improve treatment systems dedicated to reuse, and to use them on soils that are rich in organic matter, which favors pollutant degradation.

It is true that everything is being done to anticipate and prevent risks, such as optimizing treatment processes, disinfection efficiency, the application of membrane processes in the production of reused treated wastewater and the assessment of microbiological, chemical and organic contaminants.

This is indeed the key to reassuring and promoting the implementation of water reuse using emerging technologies, such as the development of new membranes adapted to wastewater, which consume less energy while providing suitable flows.

At present, the main focus is on non-drinking applications such as agricultural and landscape irrigation, industrial cooling, urban applications, etc. Water reuse could help to regulate flows and may contribute to raising groundwater levels through water reused for crop irrigation or landscaping.

Wastewater reuse guidelines are not unified around the world for the same water quality objective and the same use. The guidelines for wastewater use, published by the WHO in 2006 and 2012, are widely adopted or referenced by many countries. In addition, some countries are developing more stringent reuse standards that offer greater guarantees. For example, France's reuse standard supplements the regulation of microorganisms, including fecal enterococci, F-specific RNA phages and spores of sulfate-reducing anaerobic bacteria, compared with that of the WHO.

Treatment process	**Parameters ($mg.L^{-1}$)**						
	TSS	**BOD5**	**COD**	**Total nitrogen**	**N-NH_3**	**P-PO_4**	**Turbidity**
AS + N + P + filtration	<5	<10	<30	<5	<2	<2	<2
AS + N + P + filtration + GAC	<2	<2	<10	<5	<2	<2	<0.5
AS + N + P + filtration + O_3 + GAC	<2	<2	<10	<2	<1	<2	<0.5
AS + N + P + microfiltration	<1	<5	<20	<5	<2	<2	<0.2
MBR + N + P	<1	<5	<20	<5	<2	<2	<0.2
AS + N + MF + RO	<1	<1	<5	<0.5	<0.2	<0.1	<0.05
MBR + N + RO	<1	<1	<5	<0.5	<0.2	<0.1	<0.05

AS: activated sludge; N: nitrogen; P: phosphates, O_3: ozone; GAC: granular activated carbon filter; MF: microfiltration; RO: reverse osmosis.

Table 6.1. *Physicochemical performance of different processes*

In the United States, quality limits for reused treated wastewater differ from state to state, and the implementation of a single standard is difficult due to local conditions.

However, control and monitoring of the water produced, as well as public feedback, are necessary to assess safety. It should also be noted that cultural and religious differences shape public opinion and acceptance of wastewater reuse. Furthermore, some developing countries still implement the direct use of treated wastewater without respecting the associated standards and risks.

Implementing wastewater reuse options as part of a water resources management program requires careful planning as part of an overall management program, combined with efficient operation and management of treatment, storage and distribution facilities.

Today, there is enough technical and scientific knowledge to produce water of any quality and for any purpose. A range of technologies is available to achieve optimum effluent quality for both general and specific reuse applications.

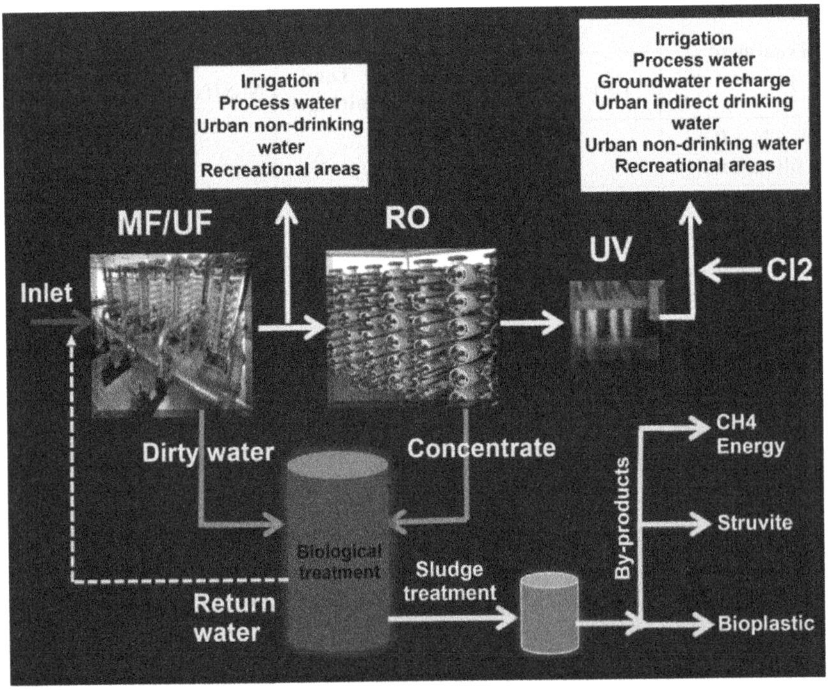

Figure 6.1. *The best way to reuse wastewater. For a color version of this figure, see www.iste.co.uk/gaid/reuse2.zip*

Sustainability is a crucial factor in water reuse, meaning that the conservation of water resources ensures that the use of technologies are reliable in their process, economically viable, institutionally recognized and socially acceptable.

6.2. What about drinking water?

Treated wastewater can be reused as drinking water in two ways: indirect drinking water and direct drinking water. Indirect drinking water is the recharging of surface water and groundwater. Direct drinking water is the most controversial type of water reuse, even if the treatments received are of a higher standard, based on more rigorous drinking water criteria. Concerns about public health, sustainability and reliability are still barriers to acceptance. As a result, only 2.3% of reclaimed water is reused in drinking water supplies.

The argument put forward by some institutions is that they respect drinking water standards and do not require any particular standards for water reused as drinking water, as long as it strictly complies with the maximum thresholds. However, concerns remain about trace organics, and the importance of public trust cannot be underestimated.

The challenges still facing membrane-based technologies for applications in groundwater recharge and the food industry are institutional, financial and cultural. The advances and limitations of various membrane processes, alone or coupled with other possible processes such as disinfection, advanced oxidation and granular activated carbon, have been highlighted by pilot tests and by a few plants operating with these processes. The latter require further research and development to ensure their reliability.

6.3. Wastewater: Is it a real alternative?

Desalination and treated wastewater reuse offer a further opportunity to avoid the complete depletion and contamination of natural water resources. Desalination would potentially reduce water stress in coastal and inland regions by producing fresh water from seawater and saline groundwater.

Wastewater reuse is an alternative way of minimizing pressure on natural freshwater resources. However, people generally accept wastewater reuse for irrigation, but not for domestic use, even if the reclaimed water meets drinking water standards.

The reuse of treated wastewater therefore offers interesting opportunities for reducing water stress and increasing water resources for a variety of uses, be it agricultural, industrial, or direct non-drinking reuse. Water reuse requires a tailor-made solution for each specific application. This is particularly true when it can replace desalinated water.

Tests at the source and advanced treatment processes combined already provide a solid basis for increased use of reused wastewater.

Parameters	Filtration	MBR	Microfiltration	Ultrafiltration	Double membrane: microfiltration-reverse osmosis
Turbidity	++	+++	+++	+++	+++
Suspended matter	++	+++	+++	+++	+++
Organic matter	–	+	–	+/–	+++
Mineral matter	–	–	–	–	+++
Organic carbon	–	+	–	+/–	+++
Color	–	++	++	++	+++
Bacteria	1 log	4 log	>5 log	>5 log	>6 log
Viruses	–	2 log	2 log	>4 log	>5 log
Parasites	1 log	5 log	>5 log	>5 log	>6 log
Helminth eggs	1 log	5 log	>5 log	>5 log	>6 log

Table 6.2. *Performance comparison between the various technologies available*

In all cases, the technologies and treatment schemes must demonstrate that the quality objectives will be achieved, while ensuring safety.

The development and marketing of new technologies opens up new opportunities for the expansion of drinking water applications, particularly in regions with water shortages. This trend will be driven by growing demand for high-quality drinking water, population growth and limited resources. This requires guaranteed access to credible scientific information for the public, decision-makers and local media in order to improve understanding of water supply in the region, as well as a cost/benefit analysis of water supply treatment options.

It has been reported that over 3,000 wastewater reuse projects in more than 60 countries have been assessed at an advanced planning stage. Most wastewater recycling systems are located in Japan (>1,800) and the United States (>800), followed by Australia (>450), Europe (>200), the Mediterranean and Middle East region (>100) and Latin America (>50). This number of projects is likely to be higher, given the rapid development of water reuse in China, India and the Middle East.

The United States stands out, accounting for more than one-fifth of the world's reused water capacity, even though it offers a national water reuse rate of just 9.7%. In areas of severe water scarcity, such as the Arab region, 18 of the 22 Arab countries were below the water poverty line of 1,000 m^3 per capita in 2014. Under the stress of shortages, 71% of wastewater collected in the Arab states was safely treated, and 21% was reused, mainly for irrigation and groundwater recharge in 2013. Singapore has a population of over 5 million, with a total water demand of 1,700,000 $m^3.day^{-1}$. Singapore has long been importing water from Malaysia to alleviate the effects of water shortages. As such, the government has launched major reuse projects.

Water-intensive regions such as China are facing severe water shortages due to rapid economic growth and urbanization, in line with their growing populations. According to China's Ministry of Ecology and Environment, 73 billion $m^3.year^{-1}$ of wastewater was produced in 2015, but only 2.9% of the total was treated for reuse. In Europe, the total volume of reused treated wastewater was 964 million m^3 per year, representing around 2.4% of treated urban wastewater effluent.

In France, there are currently around 80 wastewater reuse plants. It is likely that the drought experienced in recent summers will encourage the development of this solution. On July 25, 2022, Veolia announced its intention to deploy compact water recycling units at around 100 wastewater treatment plants, whose consumption volumes justify it (consumption of over 2,000 m^3 of drinking water and/or 5,000 m^3 taken directly from the resource). The project is expected to save 3 million m^3 of drinking water, which is equivalent to the average annual consumption of a town with a population of 180,000 people. Initially, the recycled water will be used for plant maintenance, before other urban or agricultural uses are tested, subject to the necessary permits being acquired.

For many countries, wastewater reuse will be the most effective solution to water scarcity. It is a resource that is rapidly growing. It is one of the solutions of the circular economy, which reduces the shortage of drinking water by allocating treated wastewater to agriculture and industry, thus preserving drinking water for communities.

For many scientists, reuse is an unavoidable solution if we are to limit the exponential global demand for water caused by population growth, better living conditions in developing economies, and the growth and demands of agriculture.

Developing an awareness of environmental urgency and fears of water stress, combined with mature technologies, will help to position reuse as an effective solution for guaranteeing access to water in the years to come.

It is a future solution for many parts of Africa, where, according to the UN, 75–250 million people will face major water shortages by 2030. By overcoming reluctance and dispelling certain fears, wastewater reuse will appear as a future solution for current challenges, and will see its usefulness exacerbated by the effects of climate change.

6.4. References

Baum, R., Luh, J., Bartram, J. (2013). Sanitation: A global estimate of sewerage connections without treatment and the resulting impact on MDG progress. *Environmental Science & Technology*, 47, 1994–2000.

Bixio, D., Thoeye, C., De Koning, J., Joksimovic, D., Savic, D., Wintgens, T., Melin, T. (2006). Wastewater reuse in Europe. *Desalination*, 187(1/3), 89–101.

Buyukkamacia, N. and Alkan, H.S. (2013). Public acceptance potential for reuse applications in Turkey. *Resources, Conservation and Recycling*, 80, 32–35.

Chen, Z., Ngo, H.H., Guo, W. (2013). A critical review on the end uses of recycled water. *Critical Reviews in Environmental Science and Technology*, 43, 1446–1516.

Dolnicar, S, Hurlimann, A., Grun, B. (2011). What affects public acceptance of recycled and desalinated water? *Water Research*, 45, 933–943.

Friedler, E., Lahav, O., Jizhaki, H., Lahav, T. (2006). Study of urban population attitudes towards various wastewater reuse options: Israel as a case study. *Journal of Environmental Management*, 81, 360–370.

Hartley, T.W. (2006). Public perception and participation in water reuse. *Desalination*, 6, 115–126.

de Koning, J., Bixio, D., Karabelas, A., Salgot, M., Schäfer A. (2008). Characterisation and assessment of water treatment technologies for reuse. *Desalination*, 218, 92–104.

Langeveld, J.G., Schilperoort, R.P.S., Weijers, S.R. (2013). Climate change and urban wastewater infrastructure: There is more to explore. *Journal of Hydrology*, 476, 112–119.

Major, D.C., Omojola, A., Dettinger, M., Hanson, R.T., Sanchez-Rodriguez, R. (2011). Climate change water and wastewater in cities. In *Climate Change and Cities: First Assessment Report of the Urban Climate Change Research Network*, Rosenzweig, C., Solecki, W.D., Hammer, S.A., Mehrotra, S. (eds). Cambridge University Press, Cambridge.

Mark, O., Svensson, G., König, A., Linde, J.J. (2008). Analyses and adaptation of climate change impacts on urban drainage systems. In *11th International Conference on Urban Drainage*. Urban Drainage, Edinburgh.

Mendret, J. (2023). Sécheresse : “La réutilisation des eaux usées ne peut pas résoudre à elle seule le problème”. *Le Monde* [Online]. Available at: https://www.lemonde.fr/idees/article/2023/03/07/secheresse-la-reutilisation-des-eaux-usees-ne-peut-pas-resoudre-a-elle-seule-le-probleme_6164506_3232.html.

Nasiri, F., Savage, T., Wang, R., Barawid, N., Zimmerman, J. (2013). A system dynamics approach for urban water reuse planning: A case study from the Great Lakes region. *Stochastic Environmental Research & Risk Assessment*, 27, 675–691.

Peterson, J.D., Murphy, R.R., Jin, Y., Wang, L., Nessl, M.B., Ikehata, K. (2011). Health effects associated with wastewater treatment, reuse, and disposal. *Water Environment Research*, 83, 1853–1875.

Russell, S. and Hampton, G. (2006). Challenges in understanding public responses and providing effective public consultation on water reuse. *Desalination*, 187, 215–227.

Water Research Group for 2030 (2009). Charting our water future: Economic frameworks to inform decision-making. Report, Water Research Group.

Index

A, B

C, D

K, L

M, N

O, P

R, S

T, U

V, W

Printed and bound by CPI Group (UK) Ltd, Croydon, CR0 4YY
16/09/2026
14963089-0003